The General's Playbook

The Signature Moves of History's 15 Greatest Generals

Bill Bodri

Copyright © 2025 William Bodri. All rights reserved. No part of this book may be used or reproduced in any manner whatsoever without the written permission of the publisher, except in cases of brief quotations in articles and reviews. The content of this book is for informational purposes only. The use of this book implies your acceptance of this disclaimer.

First Edition – May 2025

Top Shape Publishing LLC
1135 Terminal Way Suite 209
Reno, NV 89502
ISBN: 978-1-7370320-6-9
Library of Congress Number: 2025936903

DEDICATION

This book is dedicated to the field of military education. It strives to answer as quickly as possible the following inquiry: "What are the most effective tactics for winning battles? What winning tactics have been most often used by the greatest generals of all time?" It examines the strategies that history's greatest generals used repeatedly and successfully, distilling military wisdom into actionable insights for modern warfare. The goal is simple: to help win wars – and thus end them and their destruction – as quickly as possible.

CONTENTS

	Acknowledgments	i
1	Introduction	1
2	The Sabermetrics of War: Ranking History's Greatest Generals	5
3	Napoleon Bonaparte	13
4	Julius Caesar	22
5	Arthur Wellesley, Duke of Wellington	29
6	Takeda Shingen	36
7	Khalid Ibn al-Walid	41
8	Hannibal Barca	46
9	Ulysses S. Grant	52
10	Frederick the Great	61
11	Georgy Zhukov	67
12	Alexander the Great	73
13	Oda Nobunaga	78
14	Mustafa Kemal Atatürk	84
15	Ferdinand Foch	90
16	Douglas Haig	97
17	Augustus Caesar	104

18	Final Analysis & Thoughts	111
19	Looking Ahead: Speculations on the Future of Warfare	120
	About the Author	133

ACKNOWLEDGMENTS

Cyrus the Great, the Chinese military classics and the legendary tactician Zhuge Liang first sparked my interest in military strategy. I highly recommend the 2010 "Three Kingdoms" series (available with English subtitles on YouTube, 94 episodes) to military cadets and history enthusiasts seeking to understand his strategic brilliance.

Special appreciation goes to Ethan Arsht for his pioneering sabermetric WAR system for evaluating military commanders. I hope others will build upon his foundation to refine these rankings further, allowing us to better identify history's greatest generals and analyze their winning strategies. When commanders can achieve victory more quickly with fewer casualties and less destruction, everyone benefits—should war prove unavoidable.

1
Introduction

As a student of warfare and geopolitics, I have often wondered what specific military strategies and tactics have been historically proven as the most effective for winning battles. While military historians tend to spend a lot of time discussing a general's life history and personality (such as charisma, courage and composure), or his ability to motivate troops, conduct speedy maneuvers, trick the enemy and the like, I've always wanted a short book that just focused on the tactics that the greatest generals used to win battles in hopes of finding commonalities.

Along these lines, I have never been particularly interested in learning about a general's personality characteristics that people could not readily replicate in themselves, or revisiting basic principles of warfare that everyone would naturally know and apply. Instead, I've always sought a prioritized list of the battle-tested tactics that consistently produced victories, and that's what I hope to have produced for you.

As an example, while it is interesting to know that Napoleon was able to make enemies believe he had larger forces than actually present through strategic troop movements and disinformation, or that Caesar frequently used feigned retreats to draw enemies into disadvantageous positions, I am more interested in the actual battle strategies and tactics that each used to win battles. One answer is that Napoleon and Caesar were both masters of superior maneuverability and consistently positioned their forces to attack enemy flanks and rear positions even when outnumbered.

This was critical to their success, but the information is still too general because most every general strives for these events. "What did they do to win," I always ask myself. "What specific signature tactics did they commonly employ over and over again because I want to learn the most successful methods of history's greatest generals, first and foremost, before anything else."

Military academies traditionally teach standard warfare principles from renowned strategists like Sun Tzu, Clausewitz, Chanakya, Jomini, Hart, Boyd, etcetera – concentration of force, speed in maneuverability, deceiving the enemy, securing logistics, maintaining morale and other fundamentals such as outnumbering the enemy, striking at weaknesses, feigning retreats, keeping wars short, etc. However, I believe military cadets should first get a taste of the actual tactics and strategies of history's most successful generals before concentrating on these theoretical principles.

By initially absorbing the proven approaches of history's most victorious commanders and their stories, cadets would develop a winning mindset of possible actions that could serve them well in future battlefield situations. Only after establishing this foundation would I build upon a mental template of what has worked by then introducing the theoretical principles of winning warfare.

To clarify my approach, I would certainly teach the principles of war in any military curriculum, but I would begin by retelling the stories of what has historically worked for the most successful battlefield commanders in history, particularly when facing unfavorable odds. After students internalize these patterns, I would explore the broader principles of war, and often refer back to those stirring historical accounts to illustrate their application.

This approach requires identifying history's most successful military commanders and their winning strategies. Since human behavior has remained relatively constant throughout history, if analysis shows that numerous successful generals repeatedly employed particular strategies to win battles, doesn't it make sense to prioritize the learning of these approaches because enemies are just as likely to succumb to them? After mastering these proven tactics as a foundation, students could afterwards explore battles through additional analytical frameworks.

This is admittedly a simplified approach to a complex topic and merely reflects my preferred approach for efficiently teaching the most valuable military lessons for winning battles. Whether I am right or wrong in this approach, any aspiring commander should still ask: "What were the most successful, reliable tactics employed by history's greatest generals? Among the most successful generals, did anyone have a signature tactic that usually enabled him to win battles?"

This book attempts to answer this most basic question without unnecessary digressions on topics one should already know or surmise. If an advanced civilization from another world were to study human warfare, they would likely want to know (1) what are the principles of winning warfare in human history, the principles that guided our thinking for what wins battles, and (2) which strategies and tactics consistently succeeded throughout our history? This book specifically addresses these questions.

2
The Sabermetrics of War: Ranking History's Greatest Generals

The first step in my own educational path would be to create a list of history's finest generals. Everyone will have different opinions on absolute rankings, or even the composite collection of those who should be considered top tier commanders. To provide some objectivity, I will employ a study done by Ethan Arsht, who ranked military generals based on a baseball metric (popularized by the book *Moneyball: The Art of Winning an Unfair Game*) that estimates a baseball player's contribution to his team.

The baseball science of sabermetrics, used to determine which players a team should hire, enables you to rate a baseball player's contribution to winning games so that you can determine how valuable he is as a player. The specific metric – Wins Above

Replacement (WAR) – computes how effectively a baseball player contributes to team winnings compared to an average substitute. For example, a player with a WAR score of 5 contributed 5 additional wins to his team as compared to the expected contribution of a replacement-level player.

Instead of subjectively recounting a general's achievements, Arsht sought to use statistical battle data to more objectively determine a general's tactical fighting abilities. He adapted the Wins Above Replacement (WAR) concept to compute a score for each general in history for whom sufficient data existed.

This Arsht technique enables you to compare generals across history based on their battle performance because you can rate a general's successes *beyond what you would expect of an average general.* Arsht compared actual battle wins of every general with the victories expected of them given their military strength. He did this by computing each general's contributions above or below an average general's abilities. In other words, he compared actual battle victories to those expected of a replacement general in the same circumstances.

Arsht ranked 6,619 generals involved in 3,580 battles throughout history. For each battle, a general receives a weighted WAR score – positive for victories and negative for defeats. For example, when Napoleon fought the Russian General Mikhail Kutuzov at the Battle of Borodino, the French slightly outnumbered the Russians. Due to this slight outnumbering, the adapted sabermetric model gave a replacement general a 51% chance of victory. The WAR model awarded Napoleon a score of "1 win" for his victory, but subtracted the chance that a replacement general would have won the battle anyway, which then gave Napoleon a WAR of only .49 for his victory.

Conversely, the defeated Russian General Kutuzov received a -.49 score for the battle. By losing he achieved -1 wins, but there was a 51% chance that a replacement general would have lost anyway due to Russian forces being outnumbered. Therefore, -1 plus .51 equals -.49.

A commander who fights more battles has more opportunities to raise his WAR score by demonstrating his battle-winning capabilities. Alexander the Great, despite winning all 9 of his major battles, accumulated a lower WAR than Hannibal who fought 17 battles (winning 13, losing 2, and drawing 2).

Hannibal outscored Alexander, despite even Hannibal's assessment that Alexander was a better general, because Hannibal had 17 battles to accumulate value where Alexander won 13, lost 2 and drew 2 others to a stalemate. Hannibal's larger number of opportunities contributed to his higher WAR score.

This demonstrates that we should not argue about who was actually the better general, for that is not the purpose of this book. We are simply using a mathematical method to identify a top tier of successful generals from whom we might extract valuable lessons.

In the final rankings, Napoleon scored highest by a large margin, followed by Julius Caesar. Napoleon's exceptional ranking as the general with the highest tactical skills stems not only from winning 38 battles and losing only 5, but from overcoming unfavorable odds in 17 victories while fighting at a disadvantage in all 5 losses. Napoleon's total WAR was an incredible 23 standard deviations above the average WAR of generals in the dataset. He is definitely ranked number one for greatness above all other generals.

The victory rankings that Arsht mathematically derived are as follows:

1. Napoleon Bonaparte (43 battles, 16.703 WAR)
2. Julius Caesar (17 battles, 7.365 WAR)
3. Arthur Wellesley, Duke of Wellington (18 battles, 7.133 WAR)
4. Takeda Shingen (18 battles, 6.091 WAR)
5. Khalid Ibn al-Walid (14 battles, 5.633 WAR)
6. Hannibal Barca (17 battles, 5.489 WAR)
7. Ulysses S. Grant (16 battles, 5.023 WAR)
8. Frederick the Great (14 battles, 4.662 WAR)
9. Georgy Zhukov (10 battles, 4.595 WAR)
10. Alexander the Great (9 battles, 4.376 WAR)
11. Oda Nobunaga (11 battles, 4.229 WAR)
12. Mustafa Kemal Atatürk (11 battles, 3.582 WAR)
13. Ferdinand Foch (13 battles, 3.525 WAR)
14. Douglas Haig (12 battles, 3.517 WAR)
15. Augustus Caesar (7 battles, 3.380 WAR)

This, then, is our sabermetric list of the greatest generals in history.

Arsht's ranking will inevitably generate controversy for several reasons. To derive these numbers a database was used and some battles might not appear within that data set. Some generals certainly received credit for battles won by subordinates, some battles might be misclassified as victories or defeats, some skirmishes might be counted as battles, and of course everyone has their own opinion as to which commander was better than another.

These concerns, while valid, are ultimately irrelevant because absolute rankings aren't the point. We used the sabermetric rankings to simply identify, from countless generals, a small subset of history's most successful commanders to study their signature tactics and most effective battle strategies. *We sought a very good collection of top-tier military commanders rather than perfect rankings of who is actually better than another*, eliminating much debate over the quality of selected generals. We definitely have enough of the right commanders with high skills in this list. If we identify tactical commonalities among these generals, we will have made a significant contribution to military studies.

For completeness, there are many other generals who might also merit analysis due to their sabermetric rankings or historical importance. Additional generals meriting analysis might include:

- Alexander Suvorov
- Scipio Africanus
- Genghis Khan
- Subutai
- Belisarius
- Charlemagne
- Saladin
- Gustavus Adolphus
- Horatio Nelson
- Thutmose III
- Tariq ibn Ziyad

- Shivaji Maharaj
- Timur (Tamerlane)
- Baibars
- Merses
- George Henry Thomas
- Jan Žižka
- Ahmad Shah Durrani
- Nader Shah
- Thomas "Stonewall" Jackson
- Paul von Hindenburg
- Vo Nguyen Giap

Additionally, naval commanders would require analysis, including figures like Yi Sun-sin, Franz von Hipper, Raymond Spruance, Horatio Nelson, Michiel de Ruyter, Tōgō Heihachirō, Chester Nimitz, Themistocles, Andrew Cunningham, and Zheng He. I'm sure these results would also be interesting.

Many people are familiar with *Moneyball*, the film starring Brad Pitt that depicts how the cash-strapped Oakland A's used sabermetrics to build a competitive baseball team without expensive star players. This unconventional approach fundamentally changed their player selection process and made them a winning team.

How do teams win baseball games? By scoring runs. How do players score runs? By getting on base. Therefore, the unconventional scouting philosophy of sabermetrics holds that you must choose players who not only excel in their assigned role but *who consistently get on base*. More players on base leads to more runs scored, and scoring more runs translates into more victories.

"Getting on base" in baseball is measured by a statistic called On-Base Percentage (OBP), which measures how often a batter reaches base safely through hits, walks or being hit by a pitch. This statistic became the crucial metric for the Oakland A's, whose budget for players was half or a third as much as its rivals. The team began acquiring undervalued players to maximize this statistic *rather than focusing on traditional selection measures like batting averages* or other traditional ways for selecting players.

In sabermetric battle analysis, the key statistic might be "eliminating enemy troops" – whether by killing, capturing, or incapacitating troops, or by disrupting the enemy's ability to raise an army or send troops into battle – and therefore the most successful generals would be those who can eliminate the most troops (rather than conquer more territory), thus winning battles and wars. While this frank discussion of casualties may seem disturbing, it addresses warfare's harsh reality. It is a critical failure not to understand this.

By studying history's most successful military commanders, we can identify the strategies and tactics that most effectively neutralized enemy forces and secured victories. This is precisely the knowledge we aim to distill and acquire as efficiently as possible.

My goal is to therefore convey, as concisely as possible, the winning methods and signature tactics of history's best generals – the ways they usually won their battles. It will dispense with extraneous details that, while interesting, distract from this main issue.

The following chapters will profile the winning methods of history's greatest generals, including their signature tactics. We will not explore a commander's life history, charisma, or other personality traits. Nor will we excessively dwell on *conventional tactics that any commander would employ* in similar situations – exploiting enemy weaknesses, concentrating tactical forces at critical positions, adapting to weather conditions, psychological warfare, deception, combined-arms approaches, rapid maneuvers, leadership, or morale-building.

Instead, we will focus on core winning techniques – the primary methods and signature moves that brought battlefield success. The goal is to convey to you, in as little time as possible, the most powerful tactics or most often used techniques and strategies for winning wars without unnecessary details.

This is a simple goal, which is to answer as quickly as possible: "Which tactics and strategies should I know first and foremost for winning wars, because they've proven the most successful in human history, and which will likely remain effective in the future because human nature doesn't change?"

3
Napoleon Bonaparte (1769-1821)

According to the "Wins Above Replacement" (WAR) system of sabermetrics applied to military history, Napoleon Bonaparte achieved an unprecedented score exceeding 16 points after 43 major battles. This is so far ahead of other military commanders that it establishes him at the top of our list of winning generals even though some historians argue about his greatness.

Napoleon certainly ranks among history's greatest tactical generals, with victories in approximately 80% of his 43 major battles.[1] His ability to defeat larger or better-positioned armies stemmed from his revolutionary operational art that

---

[1] The estimates vary, but some historians cite approximately 60 engagements (including minor engagements) with Napoleon winning approximately 48-50, which is 80-83%.

combined speed, surprise, maneuver, and concentrated force tactics to annihilate enemies in decisive battles. His tactics revolutionized warfare but were built on earlier developments (such as Frederick the Great's oblique attacks) and organizational reforms like the Corps System. Napoleon's aggressive battlefield philosophy centered on destroying enemies in decisive engagements, employing innovative strategies like the central position and *manoeuvre sur les derrières*, supported by his Army Corps System that enabled him to move fast enough to achieve battlefield advantage by appearing before expected.

Napoleon demonstrated a very aggressive approach to warfare, and his battlefield philosophy centered on destroying his enemy in a single decisive confrontation instead of wearing them down over time through many clashes. Of his numerous engagements, he usually preferred offensive maneuver warfare, but fought defensively when strategically necessary, as at Leipzig (1813) and Dresden (1813).

Napoleon sought to outmaneuver his enemies and completely annihilate them rather than simply defeat or dislodge them. He strongly favored two offensive strategies for winning battles: the "central position" and "movement to the rear" (*manoeuvre sur les derrières*).

Napoleon's **central position strategy** involved positioning his army between divided enemy forces to defeat each in turn before they could unite. He would attack each force separately by concentrating on one while holding off the other. Specifically, he would divide his army into a smaller and greater part, and use his smaller part to pin one enemy, keeping them busy, while using his larger portion to crush the other. Then he would turn to the remaining foe.

A prime example of this strategy is his 1796 Italian Campaign against the Austrian and Piedmontese armies where he defeated his enemies separately before they could unite. In his 1815 Waterloo plan, Napoleon aimed to use the central position strategy by first defeating the Prussians at Ligny (June 16) while tasking Marshal Ney with delaying Wellington's Anglo-Allied forces at Quatre Bras. He aimed to prevent Wellington's unification with the Prussians since that combined force would be larger than his French army. Miscommunication and delays allowed the Prussians to regroup and reinforce Wellington, which contributed to Napoleon's ultimate defeat. Napoleon's failure at Waterloo was also partly due to miscommunication, his reliance on subordinates who lacked the initiative of his earlier marshals, and Napoleon's delayed attack allowing Prussian reinforcement, leading to defeat.

Napoleon revolutionized warfare through his **Army Corps System** that divided his Grande Armée into self-sufficient modular corps. Each corps – containing its own infantry, cavalry, artillery and support units – could operate independently for days while living off the land in fertile regions, though less effective in Russia and Spain. By splitting his army into smaller, self-sufficient units Napoleon's operational speed for advancing was quicker than that of a single massive army. These units marched separately across vast distances, often days apart, but then suddenly converged precisely for battle, as seen at Austerlitz (1805) and Jena-Auerstedt (1806). This ability to form a hammer out of separated troops (before the enemy realized the size of the total threat) and then strike with concentrated force at critical points enabled his strategy of the central position.

Throughout his campaigns, Napoleon would constantly defeat armies individually before they could combine against him. His **defeat in detail strategy**, seen in the 1796 Italian Campaign and the Six Days' Campaign (1814), involved concentrating his forces against smaller enemy detachments to prevent their unification. These smaller victories accumulated to achieve overall victory. He always used this strategy of "divide and conquer" to split enemy forces and defeat the smaller, cut-off sections one by one while preventing them from joining together to reinforce one another.

Napoleon's most often used battle tactic, the "*manoeuvre sur les derrières*" envelopment technique, used a detachment to fix enemy attention at its front while a larger force attacked its flank and rear to cut enemy supply lines and communications, as at Ulm (1805) and Austerlitz. In other words, Napoleon would hold an enemy army in place on its main line while smashing one enemy flank with overwhelming force and sending a powerful column around a flank onto its rear to force an unfavorable battle. By establishing a strategic barrage or barrier behind enemy lines this would block off all possible crossings. Napoleon would effectively isolate his opponent from its rear supply depots, sever its lines of communication, block reinforcements from arriving, and eliminate retreat options.

This tactic virtually guaranteed enemy defeat, as demonstrated at Austerlitz (1805), where Napoleon pinned the Allied center while Davout's corps struck their flank and rear, and at Ulm (1805), where Napoleon used rapid maneuvers to encircle and isolate the Austrian army, cut off Austrian supply lines, and force its surrender with minimal fighting. Basically, Napoleon's operational art relied on *rapid troop movements, isolation, and outflanking his enemies for victories*.

At Austerlitz (1805), Napoleon used **deception** by deliberately weakening his right flank to entice an Allied attack, then launched a decisive counterattack against their exposed center, securing victory. At Dresden (1813) he feigned retreat to draw Allies into a vulnerable position. He often used deception to mislead enemies by appearing weaker than he actually was in order to manipulate them to advance against seemingly vulnerable positions, encourage them to commit and overextend themselves due to overconfidence, whereupon he would launch powerful counterattacks once the enemy was committed.

As a former artillery officer, Napoleon wielded the **concentrated firepower of massed artillery** with precision. He would use mobile horse artillery to rapidly position his firepower, and would station artillery batteries at critical locations rather than distributing them across the battlefield. Napoleon massed artillery to create devastating breaches, pivotal at Friedland (1807) and Wagram (1809), where concentrated firepower supported infantry assaults followed by cavalry pursuits. Napoleon's use of artillery relied on mobility and concentration, which contrasted with the more static artillery formations of his era.

Not all battles followed this simplified pattern, but Napoleon's general battle approach typically followed a sequence along this ideal tactical framework, which varied based on circumstances:

1. Manipulate enemies into unfavorable positions
2. Fix enemy forces with frontal attacks
3. Force commitment of their main troops to the primary battle
4. Identify or create weak points in enemy lines
5. Deploy uncommitted or reserve troops (his Imperial Guard or reserve cavalry) as a late-moment shock force to enable sudden reversals when the tide of battle was turning, such as for making decisive attacks on the enemy's flanks or rear

His cavalry reserves aggressively pursued retreating enemies to prevent regrouping, often resulting in the complete destruction or surrender of the opposing force. After Jena-Auerstedt (1806), Napoleon's cavalry relentlessly pursued the retreating Prussians, preventing immediate regrouping and securing a decisive strategic advantage. His pursuits helped maximize the psychological and logistical impact of his victories.

Despite tactical brilliance, Napoleon also had flaws such as his overextension in Spain and Russia. Clausewitz criticized his strategic overreach and his miscalculation of expecting a quick Russian surrender in 1812, despite admiring his operational skill. Napoleon's reliance on living off the land, effective in Central Europe, failed in Russia (1812) due to scorched-earth tactics, vast distances, inadequate supply depots, and wrongly anticipating a quick Russian surrender, which all led to catastrophic losses.

Furthermore, as his power grew, Napoleon's earlier practice of calculating the risks in his campaigns also gave way to reckless ambition, overconfidence, and overreach. His early campaigns (before he lost his best marshals and resources) were characterized by flexible, innovative tactics. After 1807, out of necessity Napoleon increasingly relied on massed frontal assaults, as seen at Wagram (1809) and Borodino (1812), due to declining troop quality, loss of key marshals, and stronger enemy coalitions. However, he retained maneuver tactics whenever feasible (e.g., Dresden 1813).

Napoleon's fundamental principles of concentration, maneuver, timing and morale continue to influence military doctrine today. However, it must be remembered that Napoleon's focus on battlefield victories over long-term geopolitical strategy, coupled with aggressive policies like the Continental System and the Peninsular War, alienated potential allies and fueled coalitions against him. His inability to secure a lasting peace through diplomacy led to continuous wars that drained France's resources, and the empire he built lacked the institutional resilience to survive his defeat and outlast him.

Analysis of Napoleon's Tactics:

Napoleon's success stemmed from his ability to combine speed, maneuver, and concentrated force, leveraging the Corps System to outpace and outmaneuver enemies. His most frequently used tactics, employed consistently across campaigns, prioritized decisive engagements:

1. *Manoeuvre sur les Derrières* (**Flank/Rear Attack**): Used at Ulm (1805), Austerlitz (1805), and Jena-Auerstedt (1806), Friedland (1807), Marengo (1800), Eylau (1807), and Eckmühl (1809), it cut enemy supply lines and retreat routes.
2. **Central Position Strategy**: Employed at Arcole (1796), Rivoli (1797), Six Days' Campaign (1814), Ligny (1815), Marengo (1800), Dresden (1813), Castiglione (1796), Bautzen (1813), and Lützen (1813), it defeated divided enemies in detail.
3. **Corps System for Operational Speed**: Enabled rapid convergence at Austerlitz, Jena-Auerstedt, Ulm, Wagram (1809), Friedland (1807) and Eckmühl.
4. **Massed Artillery Concentration**: Pivotal at Friedland (1807), Wagram (1809), Borodino (1812), Austerlitz (1805), Ligny (1815), Eylau (1807), and Bautzen (1813), it breached enemy lines with mobile firepower.
5. **Deception and Counterattacks**: Used at Austerlitz (1805), Dresden (1813), Marengo (1800), Arcole (1796), Rivoli (1797), Castiglione (1796), and Lützen (1813), deception lured enemies into vulnerable positions.
6. **Relentless Pursuit**: Used at Jena-Auerstedt (1806), Ulm (1805), Friedland (1807), Austerlitz (1805), Wagram (1809), Eckmühl (1809), to prevent regrouping of enemy troops.

Tactical and Strategic Lessons:

- Execute *Manoeuvre sur les Derrières* for Envelopment
- Employ Central Position to Defeat in Detail
- Leverage a Corps System for Operational Speed
- Adapt Supply Systems to Local Conditions
- Concentrate Artillery for Breaching Power
- Use Deception to Manipulate Enemy Movements
- Pursue Relentlessly to Prevent Regrouping
- Balance Aggression with Strategic Restraint
- Maintain a Diplomatic Balance
- Prioritize Long-term Political Stability

4
Julius Caesar (100-44 BC)

Julius Caesar won numerous battles during the Gallic Wars and the Roman Civil War, including Alesia, Pharsalus and Thapsus. His success in the Gallic Wars (58-50 BC) frequently stemmed from **dividing and conquering** his enemies by pitting tribal factions against each other to prevent a unified Gallic resistance, a process he described in *De Bello Gallico*. This approach allowed him to fight smaller, isolated enemies rather than a larger unified force, but the strategy occasionally backfired such as when Vercingetorix united most Gallic tribes in a major rebellion, defeated at Alesia.

Caesar also preferred **outflanking and encircling** his opponents rather than frontal assaults as seen in the battles of Pharsalus (48 BC) and Alesia (52 BC). During the final days of Alesia, Caesar ordered his cavalry to ride around the outer entrenchments and surprise the enemy from the side, which forced the Gallic relief force to retreat. He often held a strong reserve and struck at enemy flanks or rear when they were overcommitted. Caesar always sought out opportunities to assault the rear of adversaries with an actual assault or a threat of attack.

Rather than any specific signature tactic, one might say that the hallmark of Caesar's winningness was his excellence in executing proven strategies with tactical flexibility that adapted to circumstances. His flexibility distinguished him from other Roman generals who tended to adhere more rigidly to traditional warfare approaches. He fused Roman discipline with mobility and aggression, training his legions for shock engagements such as at the Sabis River, Thapsus, and Pharsalus.

Caesar also demonstrated his aggressive style and tactical brilliance in his ability to execute **rapid maneuvers (*celeritas*),** his most often used tactic that enabled surprise attacks and strategic positioning. He maintained a blistering operational tempo, outmarching and outmaneuvering his enemies by regularly employing rapid forced marches (*celeritas*) that covered twice the normal Roman pace. His legions often covered 20–30 miles daily, often exceeding the typical Roman pace of 15–20 miles. His emphasis on speed is a defining trait seen in his rapid march to the Sabis River (57 BC), crossing the Rubicon (49 BC), and the Thapsus campaign. His speed disrupted enemy plans, allowing Caesar to dictate battle terms.

We must definitely attribute a portion of Caesar's victories to his swift, unexpected maneuvers.

However, Caesar's preference for rapid tactical surprise often resulted in forces too few in number to exploit the advantage of catching enemies unprepared, which sometimes brought him close to disaster (e.g., a near-disaster at Ruspina, 46 BC, where Caesar's small force was ambushed), and his boldness exposed him to unnecessary risks. He also occasionally overlooked logistical considerations, stretching his supply lines thin so that his forces suffered due to insufficient provisioning.

Each of his campaigns presented different challenges that he met with unique solutions without relying on any special signature tactic – except perhaps for his flexibility. As a general rule, however, he commonly denied his enemy food, water, or other resources to weaken them or force them into battle. At Uxellodunum, he diverted water resources to force a surrender and at Dyrrhachium (48 BC) he attempted to cut Pompey's supplies, though unsuccessful.

Caesar frequently **defied conventional military wisdom** with winter campaigns (such as his operations against the Belgae, 57 BC) when most armies remained in quarters, and adverse-weather attacks (e.g., Thapsus) when enemies least expected it.

At Pharsalus (48 BC), Caesar unexpectedly disrupted Pompey's plans by innovatively modifying the traditional Roman triple-line infantry formation by adding **a fourth line held in reserve**. This hidden fourth line surprised Pompey's cavalry-heavy wing with infantry that broke their advantage, routing their cavalry and securing victory. Caesar reverted to the traditional triple-line in other battles.

With dedicated siege and engineering experts in his army, Caesar excelled at **building and besieging**

fortifications (e.g., Alesia, Dyrrhachium), and accomplishing special engineering feats in his Gallic battles. He built siege lines, trenches and fortifications mid-battle. He constructed a 328-foot bridge across the Rhine in just ten days and would build massive siege works in record time. During the Siege of Massilia (Marseilles), he built a massive siege ramp and towers to overcome the city's defenses. At Alesia, Caesar's innovative double-walled fortifications allowed his significantly smaller force to simultaneously encircle Vercingetorix's forces while defending against relief armies – a brilliant example of defensive warfare. He was considered one of the outstanding directors of siege warfare operations in the ancient world.

Some military historians do not rank Caesar highly because he would often fall into desperate situations of his own making, but then make a genius move that turned disaster into victory. Others rank Caesar among the greatest generals based on his campaign successes involving ~50 engagements, most of which were victorious.[2] His WAR score greater than 7 after 17 battles unarguably places him in the top tier of any list of greatest generals, regardless of his actual ordinal ranking within that list, his tactical failings, or any disagreements on how such a list should be compiled.

Caesar's tactics – divide and conquer, rapid maneuvers, flanking, and siegecraft – blended Roman discipline with innovation. His practice of operational

---

[2] The sabermetric analysis only captures 17 major battles, while Adrian Goldsworthy's *Caesar: Life of a Colossus* states that Caesar fought roughly 50 engagements.

warfare created a military legacy whose influence extends to the present day. Many people know the famous victory quote attributed to Caesar: "*Veni, vidi, vici*" ("I came, I saw, I conquered"). Even today, military historians still read his *Commentarii de Bello Gallico* and *Commentarii de Bello Civili* to obtain rare firsthand insights into ancient military tactics. However, while Caesar excelled in military power, he ruled by brilliance rather than by system and lacked a sustainable vision for political stability. His consolidation of power sparked his assassination, ending the Roman Republic, which provides us with the lesson that military objectives and successes must be aligned with strategic political goals.

Analysis of Caesar's Tactics:

Caesar's success stemmed from his tactical flexibility, rapid operational tempo, and ability to adapt Roman discipline to diverse challenges. His most frequently used tactics, employed consistently across campaigns, prioritized speed and maneuver:

1. **Divide and Conquer**: Used in the Gallic Wars (e.g., against Vercingetorix's allies), Belgae Campaign (57 BC), Veneti Campaign (56 BC), Aquitanian Campaign (56 BC), Eburones Campaign (54-53 BC), Arverni Campaign (52 BC), Civil War (49-45 BC), Bibracte (58 BC), Vosges (58 BC), and Ariovistus (58 BC), it fractured enemy coalitions.

2. **Rapid Maneuvers (*Celeritas*)**: Executed at Sabis River (57 BC), Rubicon Crossing (49 BC), Alesia (52 BC), Pharsalus (48 BC), Gergovia (52 BC), Dyrrhachium (48 BC), Thapsus (46 BC), Zela (47 BC), and Bibracte (58 BC), they disrupted enemy plans.
3. **Flanking and Encirclement**: Employed at Alesia (52 BC), Pharsalus (48 BC), Thapsus (46 BC), Ruspina (46 BC), Munda (45 BC), Axona River (57 BC), and Bibracte (58 BC), they targeted enemy flanks or rear.
4. **Siegecraft and Engineering**: Mastered at Alesia, Massilia (49 BC), Uxellodunum (51 BC), Dyrrhachium (48 BC), Rhine Bridge (55 BC), and Brundisium (49 BC), and Avaricum (52 BC), it overcame fortified defenses.
5. **Resource Denial**: Used at Uxellodunum, Dyrrhachium (48 BC), Avaricum (52 BC), Massilia (49 BC), Thapsus (46 BC), and Gergovia (52 BC), it weakened enemies through starvation.
6. **Tactical Flexibility**: Demonstrated at Pharsalus (48 BC), Alesia (52 BC), Thapsus (46 BC), Ruspina (46 BC), Munda (45 BC), Sabis River (57 BC), and Zela (47 BC), Bibracte (58 BC), adapting tactics as needed distinguished Caesar from contemporaries.
7. **Unconventional Winter or Weather Campaigns:** Employed in the Belgae Campaign (57 BC), Thapsus (46 BC), Britain (55–54 BC), Eburones Campaign (54–53 BC), Arverni Campaign (52 BC), Pontus Campaign (47 BC), Ariovistus (58 BC), this leveraged surprise to gain advantage.

Tactical and Strategic Lessons:

- Employ Divide and Conquer to Fracture Enemies
- Execute Rapid Maneuvers for Surprise
- Use Flanking and Encirclement for Decisive Blows
- Master Siegecraft and Engineering for Fortified Targets
- Deny Resources to Weaken Enemies
- Adapt Tactics Flexibly to Circumstances
- Conduct Unconventional Campaigns for Surprise
- Balance Risk with Prudence
- Inspire Troops Through Your Presence

5
The Duke of Wellington, Arthur Wellesley (1769-1852)

The Duke of Wellington, who won numerous battles during the Napoleonic Wars including the decisive Battle of Waterloo, favored defensive warfare over offensive maneuvers but launched offensives when opportunities arose, as at Salamanca and Vitoria. Unlike Napoleon's aggressive, maneuver-driven style, Wellington favored patient, methodical tactics that leveraged terrain, disciplined troop formations, and strong counterattacks to defeat larger forces.

Wellington was a superb **coalition commander** who had enough diplomatic and organizational skill to integrate British, Dutch, Belgian, German, Spanish and Portuguese troops into cohesive armies against Napoleon. He leveraged their strengths – British firepower, Portuguese light infantry, Spanish guerillas – while maintaining discipline across language barriers and different troop qualities. His approach to warfare while managing these armies demonstrated a tactical philosophy that consistently defeated larger French forces.

Wellington is renowned for selecting battlefields with natural defensive features – hills, rivers, sunken lanes, or fortified farms – to anchor his lines and force enemies to attack on unfavorable terms that disrupted their formations. Selecting favorable battlefield locations, **terrain exploitation**, was his most often used strategy. It amplified his defensive strengths and helped compensate for any numerical inferiority he had in troop size.

By concealing his forces, Wellington hid the true size of his deployment, minimized its exposure to artillery, and preserved its combat power by delaying any full engagement with the enemy until it was weakened from fighting. He avoided high-risk maneuvers and preferred to let the enemy exhaust itself against his well-prepared forces. Hiding his troops behind hills or ridges enabled them to launch devastating volleys against attacking columns. At the appropriate moment, his hidden forces would suddenly emerge from cover and fire at close range in counterattacks to devastate enemy troops.

His **reverse slope defense**, a signature tactic, positioned troops behind the crest of hills or ridges, shielding them from enemy observation and artillery. This allowed his forces to remain fresh while the enemy exhausted itself advancing uphill, only to face surprise volleys and bayonet charges at close range. At Bussaco (1810), French columns crested the ridge to face devastating volleys from hidden British lines. At Waterloo (1815), the reverse slope of Mont-Saint-Jean shielded his forces, allowing them to withstand repeated French assaults until Prussian reinforcements arrived enabling a decisive general advance.

Wellington also uniquely used a **two-rank line formation** for infantry when most European armies relied on deep column troop formations. The two-line formation maximized firepower against French columns by allowing more muskets to fire simultaneously, with British soldiers trained to deliver approximately two rounds per minute versus one to two (~1-1.5) for most armies. This extra speed allowed him to create devastating volley fire that could shatter attacking columns before they reached bayonet range such as at Salamanca, where his infantry lines delivered crushing volleys that broke French columns. At Salamanca, Wellington's 52,000 troops used disciplined volleys to rout ~50,000 French, suffering ~5,200 casualties to the French's ~13,000. This firepower, combined with bayonet drills, ensured that Wellington's lines held firm under pressure.

The framework of a Wellington battle would typically run as follows: allow the enemy to attack first and then absorb their assaults. During such attacks Wellington would identify weaknesses in the enemy lines (over-extended lines or disordered formations) and then launch powerful **counterattacks** (precise opportunistic strikes) to exploit vulnerabilities that often shifted the battle's momentum.

At Salamanca (1812), he first absorbed the onslaught of French maneuvers and then launched a rapid counterattack against an overextended French flank, routing their army. At Talavera (1809) his controlled counterattacks repelled French assaults after they crested defensive positions. At Waterloo (1815), Wellington's forces held Mont-Saint-Jean until the arrival of Prussian forces bolstered a coordinated general advance. After repelling Napoleon's Imperial Guard, he ordered a decisive general advance, supported by Prussians, to rout the French army.

While Wellington preferred defensive battles, he executed bold offensives when opportunities arose as seen at Salamanca (1812), where he exploited French overextension to deliver a crushing defeat, and at Vitoria (1813) where his coordinated attack expelled French forces from Spain. In the Peninsular War (1808–1814), he also waged **attritional campaigns**, using the Lines of Torres Vedras (1810) to starve French forces while also supporting Spanish guerillas to harass supply lines.

Wellington avoided battle unless confident of success, which worked to preserve his army's strength. This cautiousness contributed to his consistent victories against numerically superior forces as it prioritized battlefield success over Napoleonic glory. Wellington's cautiousness led to limited pursuit of beaten enemies, which sometimes allowed them to escape (e.g., after Talavera). However, while Wellington was cautious to preserve his army, he pursued effectively when feasible, as at Vitoria (1813), where French forces were decisively routed and lost their baggage train. When Wellington did attack he focused on logistics to ensure that his supply lines remained secure and he avoided overextension so as not to pursue the enemy beyond sustainable limits.

Wellington's emphasis on defensive, terrain-focused tactics and line formation firepower gave him consistent advantages against numerically superior forces. He fought to win battles, not to shine with glory. His methods were the opposite of Napoleonic warfare – patient, methodical, and controlled rather than risky and dramatic – and they ultimately proved more effective. Wellington remained undefeated in major pitched battles against Napoleon's forces, culminating in his victory at Waterloo (1815), where he, alongside Prussian forces under Blücher, decisively defeated Napoleon.

Analysis of Wellington's Tactics:

Wellington's success stemmed from his patient, terrain-focused defensive tactics, disciplined firepower, and ability to integrate multinational forces. His most frequently used tactics, employed consistently across battles, prioritized defensive resilience and opportunistic offense:

1. **Terrain Exploitation (Reverse Slope Defense)**: Used at Bussaco (1810), Talavera (1809), Vimeiro (1808), Fuentes de Onoro (1811), Sorauren (1813), Waterloo (1815), Roliça (1808), Orthez (1814), it shielded troops and disrupted enemy assaults.
2. **Disciplined Two-Rank Line Formation**: Employed at Salamanca (1812), Vimeiro (1808),

Fuentes de Onoro (1811), Sorauren (1813), Assaye (1803), Waterloo (1815), Roliça (1808), and Toulouse (1814), it maximized firepower against French columns.
3. **Opportunistic Counterattacks**: Executed at Salamanca (1812), Talavera (1809), Sorauren (1813), Vimeiro (1808), Assaye (1803), Waterloo (1815), Orthez (1814), and Toulouse (1814), they exploited enemy weaknesses to shift battle momentum.
4. **Coalition Integration**: Leveraged British, Portuguese, and Spanish strengths in the Peninsular War and Waterloo for cohesive operations. Also seen at Vitoria (1813), Fuentes de Onoro (1811), Salamanca (1812), and Toulouse (1814).
5. **Attritional Campaigns**: Used at Torres Vedras (1810), Vitoria (1813), Badajoz (1811), Ciudad Rodrigo (1812), Albuera (1811), Pombal (1811), and Orthez (1814), they starved French forces.
6. **Patience with Cautious Pursuit**: Seen at Talavera (1809), Vitoria (1813), Waterloo (1815), Fuentes de Onoro (1811), Sorauren (1813), Assaye (1803), Bussaco (1810), and Salamanca (1812), this allowed French escapes but preserved his army's strength and long-term capability.

Tactical and Strategic Lessons:

- Exploit Terrain for Defensive Advantages
- Use Disciplined Line Formations for Firepower
- Launch Opportunistic Counterattacks to Shift Momentum
- Integrate Coalition Forces for Cohesive Strength

- Conduct Attritional Campaigns to Starve Enemies
- Exercise Caution to Preserve Forces
- Prioritize Logistical Security

6
Takeda Shingen (1521-1573)

Takeda Shingen, an aggressive daimyo during Japan's Sengoku period of Japan, is recognized as an incredible tactician and extraordinary military leader. He achieved remarkable battlefield success using rapid cavalry maneuvers in decisive charges to break enemy lines or execute devastating flanking attacks. His cavalry-centric approach formed the backbone of his military success, and he followed a military philosophy heavily based on Sun Tzu's *The Art of War*.

According to the *Kōyō Gunkan*, his elite mounted samurai – the Takeda *Kiba Gundan* – executed lightning-fast, devastating charges that destabilized enemies after skirmishes had disrupted their formations. Rather than engaging in individual combat, Shingen's cavalry was deployed in tight formations, using their weight and momentum to break enemy lines. Shingen would often assume defensive positions and lure enemies into attacking before unleashing his mobile cavalry offensives.

Shingen typically employed the "***Kuruma Gakari***" **(Wheel Formation)**, rotating units of cavalry and infantry in coordinated waves to maintain relentless pressure on the enemy. Shingen employed the *Kuruma Gakari* at the Battle of Mikatagahara (January 1573), rotating units to overwhelm Tokugawa's mixed forces. At Sezawa (1542), his rotation policy overwhelmed Shinano defenders. While his five battles against Uesugi Kenshin at Kawanakajima (1553–1564) were costly, the Fourth Battle showcased his tactical prowess in coordinating infantry charges with cavalry advances in a hard-fought engagement, though neither side achieved strategic dominance from the overall campaign.

While cavalry formed his specialty, Shingen **integrated all available troops** in clear, coordinated roles: *ashigaru* (foot soldiers) used spears to defend against enemy cavalry, archers provided covering fire for advancing units, and samurai infantry engaged in close combat after cavalry breached enemy formations. He coordinated their movements using drum signals, color-coded banners and fire beacons. His command structure consisted of a council of war (*Sanju-shin*) and effective delegation to trusted advisors like Yamamoto Kansuke, a key strategist, and field commanders.

Shingen's warfare philosophy – ***Furinkazan*** **("Wind, Forest, Fire, Mountain")** – emphasized swift attacks and strong defensive positioning. This doctrine translated as "swift as the wind, silent as a forest, fierce as fire, immovable as a mountain." "Wind" represented rapid movement, exemplified by his cavalry. "Forest" symbolized using terrain or concealment of movements for surprise attacks or ambushes – Shingen repeatedly used the terrain for ambushes rather than meet his enemy in the open. "Fire" signified aggressive, destructive strikes to destroy enemies. "Mountain" represented defensive fortifications, such as his fortified castles with stone foundations. Shingen's *Furinkazan* principles guided rapid-based maneuvers that often achieved the element of surprise. Shingen used *ninja* for intelligence-gathering, which supported his rapid maneuvers, and often employed terrain-based ambushes.

Shingen also employed a **woodpecker strategy** of repeatedly striking key enemy positions in a methodical fashion to erode resistance, gradually weakening their defenses before advancing. For instance, Shingen's gradual expansion in Shinano Province involved systematically besieging enemy strongholds and consolidating gains before advancing. This was very much a bite and hold strategy. He avoided overextension by building fortifications in newly conquered territories before advancing further, consolidated his gains by securing his borders before moving on, and gradually expanded his territory in this fashion.

Unlike the Duke of Wellington, who waited for enemy moves, Shingen preferred aggressive, speedy initiatives. Fast-moving cavalry was crucial to his battlefield success. Shingen's tactics – cavalry charges, terrain ambushes, *Kuruma Gakari*, troop coordination, and sieges – dominated Sengoku warfare. He struck first, engaging in prolonged skirmishes before launching concentrated assaults to break the enemy. His cavalry tactics set the standard for Japanese daimyo, and only the firearms revolution initiated by Oda Nobunaga at battles like Nagashino (1575) would later begin to challenge the dominance of cavalry-centric warfare in Japan. Shingen's military brilliance allowed him to dominate central Japan, but he did not establish a unifying political structure as Tokugawa Ieyasu did later.

Analysis of Shingen's Tactics:

Shingen's success stemmed from his aggressive, cavalry-focused approach, leveraging speed, terrain, and coordinated troop roles to outmaneuver enemies. His most frequently used tactics, employed consistently across battles, prioritized rapid offensive strikes:

1. **Rapid Cavalry Maneuvers**: Used at Mikatagahara (1573), Sezawa (1542), Mimasetoge (1569), Shiojiritoge (1548), Uedahara (1541), Fourth Battle of Kawanakajima (1561), Nagashino (1575), and Odani (1568), they broke enemy lines with elite *Kiba Gundan* charges.
2. ***Kuruma Gakari*** **(Wheel Formation)**: Employed at Mikatagahara (1573), Sezawa (1542), Fourth Battle of Kawanakajima (1561), Shiojiritoge (1548), Uedahara (1541), Odawara (1569), and Nagashino (1575), it maintained relentless pressure by rotating units of fresh troops.

3. **Terrain-Based Ambushes**: Guided by *Furinkazan's* "Forest" and "Fire" principles, they were used in Shinano Campaigns (1542–1550), Fourth Battle of Kawanakajima (1561), Mimasetoge (1569), Sezawa (1542), and Odani (1568).
4. **Coordinated Troop Roles**: Integrated *ashigaru*, archers, samurai, and cavalry at the Fourth Battle of Kawanakajima (1561), Mikatagahara (1573), Sezawa (1542), Shiojiritoge (1548), Uedahara (1541), Odawara (1569), Nagashino (1575), and Odani (1568) for cohesive assaults.
5. **Woodpecker Strategy (Gradual Sieges)**: Applied in Shinano Campaigns (1542–1550), Siege of Odawara (1569), Siege of Takatenjin Castle (1546), Siege of Hachigata Castle (1568), Siege of Katsuyama Castle (1547), and Siege of Toishi (1550), it eroded enemy strongholds methodically.

Tactical and Strategic Lessons:

- Execute Rapid Cavalry Maneuvers for Shock
- Rotate Cavalry and Infantry for Relentless Pressure
- Use Terrain-Based Ambushes for Surprise
- Coordinate Diverse Troop Roles for Synergy
- Apply Woodpecker Strategy for Gradual Erosion
- Leverage Intelligence for Strategic Advantage
- Cultivate Local Support
- Secure Territorial Gains

7
Khalid ibn al-Walid (592-642)

Khalid ibn al-Walid, a brilliant commander during Islam's early conquests, mastered aggressive mobile warfare to defeat Byzantine, Sassanid, and Arabian tribal armies. He played a key role in conquering Iraq and Syria between 632 and 636, laying the foundation for further Muslim expansion. Rarely fighting in a defensive posture, he aimed for quick military victories and avoided prolonged battles or sieges, wisely retreating when the situations were unfavorable.

Khalid's supreme tactic was **mobile warfare**, using light cavalry and rapid infantry movements to outpace enemies and strike before they could react. He maintained a highly mobile cavalry reserve positioned behind the main battle line, which was deployed at critical moments to target vulnerable points in enemy formations. This essentially amounted to an early form of rapid reaction forces.

His cavalry tactics occasionally included surprise night attacks when enemies were least prepared, but relied more on daytime flanking. He regularly conducted long-distance hit-and-run surprise raids on enemy camps and supply lines, and frequently exhausted enemies through continuous small-unit engagements before launching decisive cavalry charges.

Khalid led his army on **rapid forced marches** through seemingly impassable desert routes to achieve strategic surprise, as seen in his march from Iraq to Syria (634) via desert oases to reinforce Busra, catching the Byzantines off-guard. This required expertise at water rationing and camel logistics. A swift advance also overwhelmed Sassanid garrisons at Ullais (633). Khalid leveraged his camel-based logistics to transport infantry rapidly across desert terrain, bypassing fortified routes and enabling surprise attacks where horses proved ineffective. His rapid maneuvers and bold strikes outpaced larger enemies.

As a master of the rapid movements of mobility-based warfare, Khalid's signature tactic was **cavalry flanking**, often achieving **double envelopment** by encircling enemies from both sides. At Yarmouk (636), flanking maneuvers disrupted Byzantine lines, enabling a decisive rout. Using his light cavalry to outflank and encircle enemies was a tactic used in most major battles, making it his defining tactic. His verified tactics include cavalry flanking, double envelopment and feigned retreats, which are all well-documented.

Khalid frequently used **feigned retreats** to lure enemies into pursuit, thus disrupting their formations, followed by ambushes with hidden cavalry reserves. Khalid's disciplined troops maintained cohesion during a retreat but his enemies typically lost formation during pursuit whereby ambushes could capitalize on this disorder. Hidden reserves would strike any overextended foes in ambushes, and he would launch devastating counterattacks once the enemy overextended. At the Battle of Ajnadayn (634), feigned withdrawals baited Byzantine troops into ambushes. At the Battle of Walaja (633), a feigned retreat drew Sassanid forces into a trap where a hidden cavalry reserve enveloped them.

In multi-day battles, Khalid rotated mobile reserves to prevent fatigue while continuing the sustained pressure on enemies. For instance, at the Battle of Yarmouk (636), Khalid's forces fought over six days by using **rotating reserves**, which allowed his forces to sustain their momentum in outlasting and defeating a larger Byzantine army. At the Battle of Firaz (634), reserves defeated a Sassanid-led coalition with local Christian Arab allies.

After battle victories, Khalid relentlessly pursued defeated enemies to prevent regrouping, aiming for their total destruction. However, he offered lenient terms to surrendering cities, like Damascus (635), to conserve his forces and encourage capitulation.

Khalid's fighting tactics – cavalry flanking, rapid marches, feigned retreats, and rotating reserves – transformed Arab warfare into a systematic, mobile approach. He fought an estimated 40-50 battles and skirmishes during Islam's early expansion, though traditional sources claim up to 100, and was likely undefeated. His methods faded after his death as Muslim armies became more bureaucratic and less mobile, but he has secured a legacy as one of the greatest Muslim commanders.

Analysis of Khalid's Tactics:

Khalid's success stemmed from his mastery of mobile warfare, leveraging light cavalry, rapid infantry, and camel logistics to outmaneuver larger, less agile enemies. His tactics prioritized speed, surprise, and psychological disruption, consistently exploiting enemy vulnerabilities. The signature tactics Khalid used most often:

1. **Cavalry Flanking and Double Envelopment**: Used in nearly all major battles - Walaja (633), Yarmouk (636), Ajnadayn (634), Qadisiyyah (636), Firaz (634), Chains (633), River (633), and Marj al-Saffar (634), flanking disrupted enemy cohesion, enabling routs.
2. **Rapid Marches and Surprise Attacks**: Employed at Busra (634), Mu'tah (629), Ajnadayn (634), Yarmouk (636), Muzayyah (633), Damascus (634), Ullais (633), Homs (635), and Saniyyat (633), rapid maneuvers caught enemies unprepared, securing strategic initiative.
3. **Feigned Retreats and Ambushes**: Executed at Walaja (633), Ajnadayn (634), Qadisiyyah (636), Firaz (634), Chains (633), River (633), Marj al-

Saffar (634), and Muzayyah (633), these lured enemies into traps, capitalizing on their disorder.
4. **Rotating Mobile Reserves**: Critical at Yarmouk (636), Firaz (634), Qadisiyyah (636), and Homs (635), reserves sustained pressure, outlasting larger foes.
5. **Coordinated Troop Roles**: Seen at Yarmouk (636), Walaja (633), Qadisiyyah (636), Ajnadayn (634), Firaz (634), Muzayyah (633), Marj al-Saffar (634), and Damascus (634), cavalry, infantry and archers were integrated into a cohesive coordinated system.

Tactical and Strategic Lessons:

- Leverage Mobility for Strategic Surprise
- Master Logistics
- Employ Cavalry Flanking for Envelopment
- Use Feigned Retreats to Lure and Ambush
- Rotate Reserves to Sustain Momentum
- Pursue Defeated Enemies Relentlessly
- Offer Lenient Surrender Terms to Conserve Forces
- Integrate Small Unit Tactics

8
Hannibal Barca (247-183 BC)

Hannibal Barca, commander of Carthage, challenged Rome and won numerous battles during the Second Punic War (218-201 BC), including famous victories at Trebia, Lake Trasimene and Cannae. Hannibal's campaigns in Italy, sustained for fifteen years without a secure base, showcased his ability to outmaneuver larger Roman armies in hostile terrain, and his methods have influenced generations of commanders.

Hannibal's supreme strategy was **strategic mobility**, using rapid marches and unexpected routes to achieve surprise. His audacious crossing of the Alps in 218 BC, arriving with approximately 26,000 troops (20,000 infantry, 6,000 cavalry) and ~20-30 war elephants,[3] caught Rome unprepared, which enabled early victories. At Lake Trasimene (217 BC), he unexpectedly marched his troops through marshes to ambush ~25,000 Romans. By avoiding predictable coastal routes throughout Italy, he consistently disrupted Roman plans and forced them into reactive battles.

Hannibal's signature tactic was the **double envelopment** where he would coordinate infantry and cavalry maneuvers to encircle the enemy between two forces, thus neutralizing any numerical advantage they might have, and then destroy any troops inside the enclosure. Hannibal employed this famous double envelopment tactic at Cannae (216 BC) – now called the "Cannae maneuver" – resulting in one of Rome's most devastating defeats that trapped and destroyed an estimated 48,000-56,000 Roman soldiers, a catastrophic loss.

At Cannae, his envelopment maneuver involved intentionally weakening his center to lure Roman forces into attacking. Once the Romans committed, his center would retreat in a controlled manner, drawing the Roman forces forward. At the same time, his left and right wings would hold firm and then slowly curve inward. Hannibal would then envelop them. At Trebia (218 BC), Hannibal used a partial envelopment to trap the Romans against a river, with cavalry and hidden troops striking the rear.

---

[3] Numerical troop estimates for Hannibal are primarily derived from Polybius's *Histories* and Livy's *History of Rome*, and then adjusted according to modern scholarship.

The Romans relied on deep, frontal legions, which made them vulnerable to flanking. Hannibal's infantry would hold the line while his Numidian cavalry would dominate the flanks and exploit gaps in Roman formations before encircling the Roman infantry from behind. By attacking the Romans from multiple directions simultaneously the result was usually their complete destruction.

Hannibal also frequently employed **feigned retreats** to disorder enemy lines, followed by ambushes or counterattacks. At Lake Trasimene (217 BC), he lured ~25,000 Romans into a narrow pass, feigning a retreat before ambushing them with hidden troops along the hills. The Romans, caught in fog and confined terrain, saw losses of ~15,000 men killed or captured.

Hannibal **exploited terrain** by selecting battlefields to amplify his army's strengths – flat plains that favored cavalry flanking (Cannae) or confined spaces for trapping the Romans via ambushes (Lake Trasimene). At Trebia, his reconnaissance ensured his optimal positioning where he could trap the Romans in an unfavorable position against the river that, together with cold weather, hindered Roman mobility.

Hannibal avoided direct confrontations unless confident that his forces had significant advantages in positioning or tactical surprise. He was typically fighting against Roman deep legions, which were initially less flexible. He exploited Roman overconfidence, mistakes and predictable tactics (such as a stubborn insistence to engage in a traditional head-on battle). However, he was ultimately defeated at the Battle of Zama (202 BC) by Scipio Africanus, who used superior cavalry and adapted tactics.

Hannibal usually employed indirect approaches and unconventional tactics that countered Roman assumptions, exploiting their typical eagerness for quick victories. One of his signature tactics was to force the Romans into defensive postures, knowing their generals would feel pressured to confront him directly. He set elaborate ambushes, and during his Italian campaigns he repeatedly lured pursuing Roman armies into such traps.

Hannibal's battles aimed for total enemy destruction rather than merely to capture ground or push enemies back. He strived for complete annihilation of enemy forces whenever possible so that the Romans could never regroup or recover. By creating a military approach that repeatedly defeated larger Roman armies, his victories discredited Roman invincibility and demonstrated that a smaller force could defeat a more powerful enemy through superior tactics and strategy.

Hannibal's success during the Second Punic War stemmed from his mastery of strategic mobility, psychological deception, and combined-arms, leveraging diverse elements to outmaneuver larger Roman armies. His supreme tactics and strategies, used consistently to win battles, capitalized on terrain, surprise, and Roman vulnerabilities.

Despite Hannibal's brilliant tactics and a coalition that integrated diverse forces – Carthaginian infantry, Numidian cavalry, Gallic allies, and Spanish skirmishers – into a cohesive army, his failure to break Rome's alliances by rallying Italian city-states to his cause, his inability to secure long-term

Carthaginian support for significant reinforcements or naval support, and his failure to convert victories into strategic gains, prevented him from dismantling Roman power despite his battlefield success. He was tactically brilliant, but in the end his strategic goals were thwarted by Rome's resilient alliances and Carthage's limited support.

Analysis of Hannibal's Tactics:

Hannibal's success stemmed from his ability to outmaneuver larger Roman armies through speed, deception, and combined-arms coordination. His most frequently used tactics, employed consistently across battles, prioritized exploiting Roman predictability and terrain:

1. **Double Envelopment (Cannae Maneuver)**: Used at Cannae (216 BC), Ticinus (218 BC) and partially at Trebia (218 BC), it encircled enemies for annihilation.
2. **Strategic Mobility and Surprise**: Rapid marches - Alps Crossing (218 BC), Trasimene (217 BC), Cannae (216 BC), Geronium (217 BC), and Ticinus (218 BC), they caught the Romans unprepared.
3. **Feigned Retreats and Ambushes**: Executed at Trasimene (217 BC), Trebia (218 BC), Geronium (217 BC), and Cannae (216 BC), these disordered Roman formations for counterattacks.
4. **Terrain Exploitation**: Selected battlefields (e.g., Cannae (216 BC), Trasimene (217 BC), Trebia

(218 BC), Ticinus (218 BC), Geronium (217 BC)) amplified cavalry and ambush effectiveness.
5. **Combined-Arms Coordination**: Integrated infantry, Numidian cavalry, and skirmishers for cohesive assaults, seen in all major battles especially at Cannae (216 BC), Trebia (218 BC), Trasimene (217 BC), Ticinus (218 BC), Zama (202 BC), and Geronium (217 BC).

Tactical and Strategic Lessons:

- Execute Double Envelopment to Annihilate Enemies
- Leverage Strategic Mobility for Surprise
- Use Feigned Retreats to Disorder and Ambush
- Exploit Terrain to Amplify Strengths
- Exploit Psychological Warfare
- Integrate Diverse Forces
- Coordinate Combined-arms for Synergy
- Exploit Enemy Predictability and Overconfidence
- Aim for Total Enemy Destruction

9
Ulysses S. Grant (1822-1885)

Ulysses S. Grant is known for his relentless, offensive approach in warfare that won numerous battles during the American Civil War, including Vicksburg and the Appomattox Campaign. Rising from the Union's most successful general to commander of all Union armies from 1864, Grant's pragmatic strategy of continuous pressure on Confederate forces, focusing on attrition and infrastructure rather than rigid battle formations, diverged from the Napoleonic tactics that influenced many other Civil War generals. Grant's hallmark was relentless offensives to exhaust Confederate resources, exploiting Union manpower and industrial advantages to win the Civil War. He is sometimes considered America's first modern military commander.

Grant's core approach employed both **flanking** (e.g., Vicksburg, Chattanooga) and **envelopment maneuvers** but he also employed frontal assaults, sometimes costly ones (e.g., Cold Harbor in 1864) that hammered at the enemy's center. Grant combined an initiative to keep advancing regardless of setbacks with tactics of attrition, siege warfare and a "total war" focus that aimed to destroy the Confederacy's ability and will to continue fighting. His strategy was to exploit the North's superiority in men and materials, attack the Confederacy on many fronts, and keep up continuous pressure until the South had become so weakened that it would be forced to surrender.

Grant absolutely favored forward movement over defensive positions. While he initially launched direct attacks on fortified Confederate positions that resulted in heavy losses, as Confederate forces entrenched, Grant increasingly favored flanking maneuvers, though he still used frontal assaults when necessary, as at Spotsylvania which combined frontal assaults with limited flanking attempts against Lee's entrenchments. During the Vicksburg Campaign, he avoided the city's fortified defenses, bypassing the strong Confederate position using inland marches, feints, and a bold flanking maneuver by crossing the Mississippi River south of the city. This campaign showcased Grant's logistical skill because he needed to sustain his army without a supply line for weeks, a daring move for the time. At Chattanooga (1863), Grant flanked Confederate positions at Lookout Mountain.

Many of Grant's successes were due to superior manpower and resources rather than tactical brilliance alone. However, as a point of interest Grant's sabermetric WAR score of 5.023 after 16 battles firmly beats Robert E. Lee's WAR score of -1.994 after 27 battles, suggesting Grant as the better general overall according to *this* rating scheme, which does not account for qualitative factors like logistics or circumstances. It is absolutely true that Grant's strategy leveraged Union resources to outlast Lee, whose tactical brilliance was constrained by Confederate limitations. A further notable statistic is that in all the battles he fought Grant incurred ~25% losses compared to Lee's ~46% casualties, though Grant's larger armies sustained higher absolute casualties.[4]

A comparison of the two generals across multiple dimensions is controversial using any scheme, and I am supplying these figures only because most everyone is curious about the sabermetric scores of these two generals. Remember that we are not using the sabermetric approach to rank generals but simply to identify, from a list of 6,619 military commanders, a top subset we might analyze to discover the best or most common successful tactics and winning strategies for warfare.

Grant's **attrition warfare** strategy involved attacking repeatedly with persistence to wear down Confederate forces, and you can argue that attrition and continuous pressure was Grant's signature tactic. Grant's Overland Campaign (1864) exemplified this

---

[4] Estimates from Bruce Catton's *Grant Moves South* and *Grant Takes Command*.

approach, with battles like the Wilderness, Spotsylvania, and Cold Harbor designed to bleed Lee's Army, even at high Union cost. His strategy exploited the Union's manpower and industrial advantages. He aimed not only to inflict heavy casualties and exhaust Confederate resources so it could not continue fighting, but to cause the Confederacy to make tactical errors or overextend during the conflict. Attrition was not Grant's sole strategy but a deliberate choice in response to facing Lee's entrenched army.

The attrition strategy of continuous fighting was a bloody, grinding style of warfare that cost him heavy casualties, but Grant calculated that the Union could replace soldiers and materials faster than the South. Wearing it down would ultimately favor the Union in the long-run, but at a high cost Grant was willing to accept if the casualties disproportionately weakened the Confederacy. During the Overland Campaign, for example, he continued attacking despite heavy casualties because Lee could not replace the comparable losses, and because the attack prevented Lee from reinforcing other fronts.

Grant's relentless pressure kept Confederate forces off-balance and unable to regroup effectively. After each battle, he refused to pause operations but would **pursue the enemy relentlessly** to either destroy them or force surrender. He would just continue moving forward, capitalizing on Confederate disarray, in an aggressive pursuit of his strategic objectives. After the Battle of Fort Donelson, for instance, he immediately pursued fleeing Confederates rather than consolidate his position, and secured Union control of western Tennessee.

Grant complemented his basic approach by authorizing and coordinating **strategic raids** into enemy territory targeting critical infrastructure like railroads, bridges, factories, and supply depots. As an example, Grant authorized Sherman's March to the Sea, which targeted Georgia's infrastructure and morale. The Siege of Petersburg (1864–1865) cut Confederate rail lines, starving Lee's army and forcing surrender. By destroying the South's war-making capacity, Grant aimed at removing its resources and breaking its will to fight.

As a form of psychological warfare, Grant also used ultimatums, propaganda, and symbolic victories to demoralize Confederate forces, inspire Union troops, and sway civilian opinion. His reputation for relentless pursuit and harsh terms, like "unconditional surrender," intimidated enemies and often prompted capitulation. This tactic played a definite role in his victories, particularly at Appomattox, because his psychological blows helped break Confederate morale.

Grant masterfully employed **joint Army-Navy operations**, using riverine gunboats at Fort Henry to open the Tennessee River as a Union supply line, at Fort Donelson to access the Cumberland River, and at Vicksburg to support infantry advances. Under cover of darkness, at Vicksburg Grant flanked its batteries in a calculated high-risk operation of using gunboats to move men and supplies south to avoid assaults.

Grant readily **embraced new technologies** such as using railroads for troop movements and supply, telegraphs for communication, and ironclad ships for naval support. Grant's use of railroads and telegraphs enabled his unprecedented coordination of multiple armies, a hallmark of his 1864–65 strategy.

Through **siege operations** (by building miles of trenches to contain and bleed the enemy) he applied continuous pressure against enemy troops and cut off supply lines until Confederate forces were forced to either surrender or be destroyed. As an example, the 47-day Siege of Vicksburg isolated Confederate forces until they surrendered, giving the Union control of the entire Mississippi River and splitting the Confederacy into two.

Grant embraced "**total war**," aiming to not only destroy Confederate armies but also the South's military, economic and logistical infrastructure to cripple its war-making capacity. The goal was not only to break the South's will to fight but to make it impossible for the Confederacy to wage war due to a lack of men and war-sustaining resources.

This approach counteracted the Confederate strategy for winning by "not losing" – simply holding out until the Northern will to fight collapsed due to weariness. Grant moved quickly and leveraged the Union's superior manpower and resources to overwhelm the Confederacy and outlast it as the North was better positioned to sustain a war of attrition. His victories were not achieved through the tactical brilliance of complex maneuvers but by leveraging a combination of attrition, logistical superiority, and coordination, complemented by tactical ingenuity in campaigns like Vicksburg and Chattanooga. His total war approach and strategy of continuous pressure was well ahead of the military doctrine of his time.

Grant's broad strategic vision was to wear down the Confederacy through *coordinated sequences of battles* to capture states by cutting off their forces from each other's resources. For Grant, winning was not about singular battle campaigns but about *how to make multiple armies do multiple things to accomplish one big goal* – a final victory. He **synchronized multiple armies** across vast distances to coordinate war on a national scale. In 1864, he synchronized five separate Union armies across different theaters – the Army of the Potomac (Meade), Sherman's armies in Georgia, Banks in Louisiana, Sigel in the Shenandoah Valley, and Butler near Richmond.

His simultaneous assaults on multiple fronts was a novel strategy that prevented the Confederacy from shifting forces for strategic concentration, and his total war approach of attrition, continuous pressure, flanking, leveraging resource advantages, and coordinated operations ultimately secured Union victory. One might argue that Grant was the general primarily responsible for winning the Civil War.

Analysis of Grant's Tactics:

Grant's success stemmed from his pragmatic use of Union manpower and its industrial advantages, combined with relentless pressure and logistical innovation. His most frequently used tactics, employed consistently across campaigns, prioritized wearing down Confederate forces and infrastructure:

THE GENERAL'S PLAYBOOK

1. **Attrition and Continuous Pressure**: Used in the Overland Campaign (1864), Vicksburg Campaign (1863), Shiloh (1862), Appomattox Campaign (1865), Iuka (1862), Cold Harbor (1864) and at Petersburg (1864–1865), it bled Confederate resources.
2. **Flanking and Envelopment Maneuvers**: Executed at Vicksburg (1863), Chattanooga (1863), Fort Donelson (1862), Champion Hill (1863), Iuka (1862), and Spotsylvania (1864), these bypassed fortified defenses.
3. **Total War and Infrastructure Destruction**: Seen in Sherman's March (1864), Petersburg Siege (1864-1865), Shenandoah Valley Campaign (1864), and Atlanta Campaign (1864), it crippled Confederate logistical and economic infrastructure.
4. **Joint Army-Navy Operations**: Leveraged at Vicksburg (1863), Shiloh (1862), Mobile Bay (1864), and Forts Henry/Donelson (1862), and Port Hudson (1863), integrating riverine support.
5. **Multi-Front Coordination**: Implemented in 1864–1865 Campaign, Vicksburg-Chattanooga (1863), Atlanta Campaign (1864), Appomattox (1865), and Shenandoah Valley Campaign (1864), synchronizing multiple armies to prevent Confederate reinforcement and stretch their resources.
6. **Relentless Pursuit**: Used at Fort Donelson (1862), Appomattox (1865), Shiloh (1862), Champion Hill (1863), and Chattanooga (1863), to prevent Confederate regrouping.
7. **Siege Warfare:** Used at Vicksburg (1863), Petersburg (1864-1865), Port Hudson (1863) and

Corinth (1862), to starve forces and force surrender.

Tactical and Strategic Lessons:

- Employ Attrition to Exhaust Enemy Resources
- Execute Flanking Maneuvers to Bypass Defenses
- Target Infrastructure for Total War
- Integrate Joint Army-Navy Operations
- Synchronize Multi-Front Operations
- Pursue Relentlessly Post-Battle
- Use Psychological Pressure
- Embrace Technology for Coordination
- Conduct Siege Warfare to Starve Enemies
- Balance Casualties With Strategic Goals
- Foster Subordinate Autonomy

10
Frederick the Great (1712-1786)

Frederick II of Prussia, known as Frederick the Great, almost always fought outnumbered during the Seven Years' War, but he won numerous battles, including Rossbach and Leuthen where he defeated Austria despite almost 2:1 odds. He developed distinctive battlefield approaches that allowed his smaller kingdom to defeat the combined might of Austria, France, and Russia.

Frederick's success stemmed more from offensive actions than from defensive warfare. He believed in conducting war based on fixed principles that could be understood by studying past conflicts. Consequently, he became an ardent student of military history, studying the campaigns of Caesar, Gustavus Adolphus, and other successful commanders in order to apply what they had learned.

Frederick improved artillery mobility with horse teams and lighter guns that could rapidly reposition during battles. Typically, he concentrated his artillery batteries at critical points rather than spreading them thinly across enemy lines. He positioned them to soften enemy positions before infantry assaults (as at Leuthen), and to achieve decisive breakthroughs.

Frederick also relied on highly disciplined infantry drill-trained to fire 3–4 rounds per minute under ideal conditions, but in combat, approximately 2 rounds per minute was typical compared to ~1–1.5 for Austrian or French infantry. In short, his firepower outpaced and overwhelmed slower enemies, thus reducing his own casualties during attacks and retreats. At Leuthen (1757) his infantry volleys shattered Austrian lines after a flank attack, and at Hohenfriedberg (1745) his rapid infantry fire supported cavalry charges.

Frederick's signature tactic was the "**Oblique Order**" (*Schräge Schlachtordnung*), which allowed him to defeat larger forces repeatedly. It was an angle of attack used in most major victories, and represents his tactical legacy. The oblique order involved reinforcing one of his own flanks so that this strengthened force could strike the enemy's flank at an angle while his other flank engaged the enemy minimally, remaining in a supporting role.

This tactic concentrated forces to overwhelm a specific sector of the enemy's flank while pinning down the enemy's main army with a smaller wing so that it could not effectively redistribute its forces. By concentrating forces, he achieved local superiority at the point of attack despite his overall numerical inferiority. Troops then advanced in echelon (at an angle) to allow successive waves of attacks to roll up the enemy line after breaching the targeted flank. At the Battle of Leuthen (1757), he used this oblique order formation to defeat an Austrian army twice his size. At Hohenfriedberg (1745), an oblique attack routed Austrian and Saxon forces.

Frederick occasionally adopted defensive postures to lure enemies into unfavorable positions and expend their energy before counterattacking, as at Hohenfriedberg (1745), but he preferred offensive maneuvers. However, he was also known for launching preemptive strikes, especially in the Silesian Wars and at the Invasion of Saxony (1756). Known for his rapid maneuvering, he would strike first in battle to set the terms of the conflict as this would put him in the driver's seat.

The foundation of Frederick's tactical system was a **rigorous training program of standardized drills** that produced unprecedented military discipline. This rigorous drilling allowed his troops to perform complex maneuvers impossible for other armies. His cavalry, trained to charge at full gallop with swords (avoiding firearms to maximize the impact), delivered shock action to break enemy lines or exploit breaches. At Hohenfriedberg (1745), his cavalry smashed Austrian-Saxon flanks. At Rossbach (1757), charges routed French-Allied troops, exploiting breaches with speed and impact.

Frederick trained his army to **march faster and more cohesively** than any other European army. He then used exceptionally fast marches to concentrate forces faster than his opponents, and to outflank and encircle enemy forces. Rapid maneuvers allowed 22,000 Prussians to outflank 42,000 French-Allied troops at Rossbach, and a preemptive strike secured Prussian initiative at the Invasion of Saxony (1756) during the outset of the Seven Years' War.

Frederick always sought decisive victories where he lost the fewest troops possible. He avoided reckless pursuits that risked his army, never placing it in danger of annihilation, as seen in his orderly retreat at Kolin (1757) after a failed attack. Unlike many commanders of his era, he conducted orderly withdrawals when necessary. One could say that he won by surviving rather than by always achieving decisive breakthroughs.

Frederick's tactics – oblique order, rapid maneuver, mobile artillery, disciplined infantry, and shock cavalry – overcame his numerical inferiority in battles. His rigorous training produced an army capable of complex maneuvers, as seen at Leuthen and Rossbach. His manuals, like *Instructions for His Generals*, influenced military thought into the 19th century. He demonstrated that smaller armies could defeat larger ones by concentrating inferior troop numbers rapidly and then focusing on weak spots in the enemy line. However, Frederick's aggressive wars strained Prussia's diplomacy but secured Silesia, cementing its status as a great power.

Analysis of Frederick's Tactics:

Frederick's success stemmed from his ability to maximize a smaller army's effectiveness through discipline, speed, and tactical innovation. His most frequently used tactics, employed consistently across his battles, prioritized offensive maneuvers and concentrated force:

1. **Elastic Defense with Counterattacks**: Employed at Leuthen (1757), Rossbach (1757), Zorndorf (1758), Soor (1745), Hohenfriedberg (1745), Kolin (1757), Torgau (1760), Liegnitz (1760), Mollwitz (1741), Prague (1757) and Burkersdorf (1762), assaults were absorbed before offensive counterstrikes.
2. **Oblique Order, Flanking and Envelope Maneuvers**: Used at Leuthen (1757), Rossbach (1757), Kolin (1757), Soor (1745), Hohenfriedberg (1745), Zorndorf (1758), Kunersdorf (1759), Liegnitz (1760), Torgau (1760), Mollwitz (1741), Prague (1757) and Burkersdorf (1762), it overwhelmed enemy flanks for decisive breakthroughs.
3. **Rapid Maneuvers and Preemptive Strikes (Aggressive Offense)**: Executed at Saxony (1756), Leuthen (1757), Rossbach (1757), Zorndorf (1758), Soor (1745), Hohenfriedberg (1745), Kolin (1757), Torgau (1760), Mollwitz (1741), Chotusitz (1742), Liegnitz (1760), and Prague (1757), these seized the initiative, disrupted enemy plans and outflanked enemies.
4. **Combined-Arms Coordination**: Leuthen (1757), Rossbach (1757), Zorndorf (1758), Kolin (1757), Soor (1745), Hohenfriedberg (1745), Torgau (1760), Mollwitz (1741), Liegnitz (1760), Chotusitz (1742), Hochkirch (1758), and Burkersdorf (1762), integration of cavalry, infantry and artillery maximized battlefield impact.

5. **Disciplined Infantry Firepower**: Leveraged at Leuthen (1757), Hohenfriedberg (1745), Rossbach (1757), Zorndorf (1758), Torgau (1760), and Prague (1757), rapid volleys shattered enemy lines.
6. **Mobile Artillery Concentration**: Deployed at Leuthen (1757), Rossbach (1757), Zorndorf (1758), Torgau (1760), and Burkersdorf (1762), it softened targets for infantry assaults.
7. **Rapid Maneuvers and Surprise Attacks**: Used at Leuthen (1757), Rossbach (1757), Zorndorf (1758), Soor (1745), Hohenfriedberg (1745), Kolin (1757), Torgau (1760), Mollwitz (1741), Chotusitz (1742), Liegnitz (1760) and Hochkirch (1758).

Tactical and Strategic Lessons:

- Study Historical Precedents
- Maintain the Strategic Initiative
- Employ the Oblique Order to Overwhelm Enemy Flanks
- Execute Rapid Maneuvers for Preemptive Strikes
- Leverage Disciplined Firepower for Superiority
- Concentrate Mobile Artillery for Breakthroughs
- Deploy Shock Cavalry for Decisive Impact
- Disciplined Training for Complex Maneuvers
- Conduct Orderly Withdrawals to Preserve Forces

11
Georgy Zhukov (1896-1974)

Marshal Georgy Zhukov was an aggressive, offensive-oriented Soviet general who fought decisive large-scale battles during World War II such as the Battle of Stalingrad, Battle of Kursk, Operation Bagration and Berlin Offensive. His success in World War II stemmed from his ability to orchestrate massive offensive operations where he leveraged Soviet manpower, industrial capacity, and mechanized forces to overwhelm German armies.

Zhukov implemented the Soviet doctrine of **"Deep Operations"** that employed multiple echelons attacking in waves to penetrate enemy lines, strike deep into rear areas with fast-moving forces, and collapse enemy cohesion. Zhukov would simultaneously attack across the entire depth of enemy defenses while working to create specific corridors of advance for mechanized assaults that would target the enemy's rear. In the Berlin Offensive, over 40,000 guns (artillery pieces, mortars and rocket launchers) opened the assault. He concentrated artillery fire on narrow fronts to create breakthroughs (using layered artillery barrages timed down to the minute) before pushing tanks and infantry through those gaps. At Bagration (1944), artillery shattered German lines, facilitating deep penetrations.

His signature tactic was **double envelopment, a large-scale encirclement** where Soviet forces would use simultaneous attacks from multiple angles to encircle enemy forces in a "cauldron" (*Kesselschlacht*) that prevented retreat, ensuring the enemy's total destruction. This was central to his major victories, and defined his operational style. His Operation Uranus successfully encircled the German 6th Army (~250,000-270,000 troops) at Stalingrad, leading to its destruction with ~90,000 surviving to surrender. In Operation Bagration, giant flanking thrusts destroyed German divisions (~350,000-450,000 troops) via multi-directional assaults. When necessary, Zhukov would employ costly frontal attacks to pin down German forces while other sectors advanced or regrouped. At Seelow Heights (1945), Zhukov used frontal assaults to break German defenses, but the heavy losses prompted flanking maneuvers and additional artillery to breach defenses.

Zhukov often used **operational deception** (*maskirovka*) to mislead enemies with elaborate dummy installations, false radio traffic, and feigned troop concentrations before battles. He would conduct conspicuous preparations in non-critical areas to mislead German intelligence while assembling reserves outside enemy observation. At Kursk (1943) and Stalingrad (1942), German forces were caught off-guard by the actual location and timing of Soviet attacks because of his deceptions.

Despite the use of misdirection, Zhukov primarily relied on overwhelming firepower and massed infantry assaults to win battles. He often used night operations to gain surprise, as at Stalingrad and Bagration. He also timed offensives to exploit the Russian weather conditions, such as launching a winter offense when German mobility was reduced.

Zhukov **coordinated massive offensives across different fronts** with over one million troops as in Bagration and Berlin. Because he typically relied on overwhelming troop numbers and massed offensives one should argue that some of his success must be attributed to sheer numerical superiority and Russia's industrial capabilities rather than just tactical brilliance and superb operational planning. Zhukov rarely used rapid maneuvers in his battles but instead relied on overwhelming firepower and massed infantry assaults.

Zhukov commonly fortified his defenses to weaken German offensives before counterattacking, such as at Kursk where Soviet defenses blunted Germany's Operation Citadel and then drove back German Panzers, but his primary focus was aggressive, large-scale offensives. He would allow enemies to attack first, wear them down through strong defensive positions, and then launch powerful counterattacks with superior forces when the enemy was exhausted. He maintained significant reserves ready to commit to battle when the enemy became worn out or overextended, and deployed them at critical moments to add reinforcements, strike at enemy weak points with concentrated force, or to exploit breakthroughs.

Zhukov's primary tactics – double envelopment, deep operations, *maskirovka*, massive artillery barrages, and counterattacks – leveraged Soviet resources, but it was his great ability as an operational commander that enabled him to coordinate massive forces with precision in large-scale battles. His military success stemmed from this mastery of large-scale offensive operations to overwhelm German armies. His tanks and mechanized corps would rapidly push deep into enemy rear areas to create operational encirclements. Zhukov's offensives combined the pressure of attritional warfare to wear down German capabilities over time with these skillful encirclements to destroy German armies, but it cost him massive casualties.

Analysis of Zhukov's Tactics:

Zhukov's success stemmed from his ability to orchestrate massive coordinated offensives, leveraging Soviet resources and deception to outmaneuver and destroy German armies. His most frequently used tactics, employed consistently across major battles, prioritized overwhelming force and encirclement:

1. **Double Envelopment (*Kesselschlacht*)**: Used at Stalingrad (1942–1943), Bagration (1944), Vistula-Oder Offensive (1945), Khalkhin Gol (1939), Smolensk (1943), Korsun-Cherkassy

(1944), Kiev (1943), Berlin (1945), Operation Uranus (1942), and the Jassy-Kishinev Offensive (1944), it encircled and annihilated enemy forces.
2. **Deep Operations**: Executed at Bagration (1944), Berlin (1945), Vistula-Oder Offensive (1945), Operation Kutuzov (1943), and the Jassy-Kishinev Offensive (1944), these penetrated enemy rear areas to collapse defenses.
3. **Operational Deception (*Maskirovka*) and Feigned Retreats**: Employed at Kursk (1943), Bagration (1944), Moscow (1941–1942), Khalkhin Gol (1939), Smolensk (1943), Korsun-Cherkassy (1944), Rzhev-Vyazma (1942), Stalingrad (1942–1943), Operation Uranus (1942), and Jassy-Kishinev Offensive (1944), it misled enemies about Soviet intentions.
4. **Concentrated Artillery Barrages**: Used in all major offensives (e.g., Bagration (1944), Berlin (1945), Kursk (1943), Stalingrad (1942–1943), Khalkhin Gol (1939), Smolensk (1943), Seelow Heights (1945), Kiev (1943), Operation Uranus (1942), and Operation Kutuzov (1943)), they created breakthroughs for mechanized assaults.
5. **Coordinated Troop Roles**: Integrated infantry, artillery, armor and air support into cohesive formations, seen at Kursk (1943), Stalingrad (1942-1943), Bagration (1944), Berlin (1945), Khalkhin Gol (1939), Smolensk (1943), Korsun-Cherkassy (1944), Kiev (1943), and Operation Uranus (1942), Jassy-Kishinev Offensive (1944).
6. **Defensive Preparation with Counterattacks**: Applied at Kursk (1943), Moscow (1941-1942), Leningrad (1941-1944), Stalingrad (1942-1943), Khalkhin Gol (1939), Smolensk (1941), Rzhev-

Vyazma (1943), Kiev (1943), and Operation Kutuzov (1943), it blunted enemy offensives before decisive counterstrikes.

Tactical and Strategic Lessons:

- Execute Double Envelopment for Annihilation
- Implement Deep Operations to Collapse Rear Areas
- Use Operational Deception (*Maskirovka*) for Surprise
- Leverage Psychological Warfare
- Concentrate Artillery for Breakthroughs
- Fortify Defenses Before Counterattacking
- Exploit Environmental Conditions for Advantage
- Maintain Reserves for Decisive Moments
- Coordinate Multi-Echelon Attacks

12
Alexander the Great (356-323 BC)

Alexander the Great, undefeated in major pitched battles, conquered the Persian Empire and beyond with lightning campaigns across three continents that showcased tactical genius. At Gaugamela (331 BC) he defeated a Persian army roughly two times larger than his own, and he overcame enemies like Darius III and Porus through innovative combined-arms strategies. Alexander typically found some way to bypass an opponent's strength to create or exploit any vulnerability in their flanks and rear.

Alexander's signature tactic and the core of the Macedonian military approach was a "**hammer and anvil attack**" to pin the enemy's front while striking its flanks. The Macedonian phalanx and *hypaspists* (elite infantry), armed with long pikes, formed a dense "anvil" to pin the enemy's front and absorb attacks. His heavy cavalry (*hetairoi*), often led personally, acted as a "hammer" to strike enemy flanks or rear.

Alexander would also target the opposing commander to create panic and confusion in his troops, as happened when his cavalry targeted Darius III forcing his flight that triggered Persian collapse. In other battles, he captured Porus at Hydaspes and targeted Ariobarzanes at the Persian Gates, breaking resistance. Alexander used the hammer and anvil in all major pitched battles (Granicus, Issus, Gaugamela, Hydaspes), making it his hallmark strategy. During such battles, Alexander himself typically led the cavalry on the right flank while his infantry held the center.

Alexander's victories were achieved by **integrating phalanx, cavalry, *hypaspists*, and light infantry** into a cohesive system, each unit supporting the others. Light infantry (e.g., Agrianians) harassed enemies, while his cavalry and phalanx delivered decisive crushing blows. At Gaugamela (331 BC), his light infantry screened the phalanx, enabling cavalry strikes while at Hydaspes (326 BC) his light infantry targeted elephant drivers, supporting cavalry and phalanx flanking assaults. He employed this combined-arms strategy consistently, whose coordination allowed him to remain undefeated on the battlefield although smaller engagements, like the siege of Multan (325 BC), involved heavy Macedonian casualties and wounded Alexander severely.

Alexander prioritized **flanking maneuvers**, attacking enemy sides or rear to create confusion and causing them to fight on multiple fronts. He often used **oblique attacks**, advancing one wing while refusing the other to achieve local superiority and unbalance the enemy line. At the Battle of Chaeronea (338 BC), under Philip II, Alexander's cavalry exploited an oblique attack (striking the enemy's line from unexpected angles) to break the Theban line. At Issus (333 BC), Alexander led his cavalry across the Pinarus River to outflank the Persian right, targeting Darius's vulnerable center. At Hydaspes (326 BC), Alexander outflanked Porus's forces, using light infantry to harass elephants and cavalry to break the Indian line.

Although the genius of Alexander lay in coordinating combined-arms – phalanx, cavalry, *hypaspists* and light infantry – as a cohesive system, he also excelled in **siege warfare**, using engineering like causeways and torsion catapults to overcome fortified defenses. At Tyre (332 BC), he built a causeway to breach the island fortress others deemed impenetrable. At Sogdian Rock (327 BC), elite climbers scaled cliffs, forcing the enemy's surrender.

Alexander also mastered logistical planning, establishing supply chains, depots, and local alliances for the daunting task of sustaining his rapid campaigns across vast distances. His ability to maintain armies in hostile terrain, from Persia to India, depended on his supply chain management skills which ensured operational continuity.

Alexander's tactics – hammer and anvil, flanking, the integration of forces for combined-arms, commander targeting, and siege innovation – secured an undefeated record in major battles. Despite great battlefield victories, Alexander can be criticized for taking unnecessary risks through personal exposure, such as leading cavalry charges himself. While inspiring, his need to lead from the front almost got him killed multiple times, and he suffered severe wounds in battle that recklessly endangered the continuity of his leadership.

His army fragmented after his death due to the lack of a clear successor, showing that his military successes were more dependent on his personal leadership than on a sustainable system, but his military legacy endures as a benchmark for tactical brilliance. Even so, his failure to institutionalize a long-term governance structure led to the rapid collapse of his empire.

Analysis of Alexander's Tactics:

Alexander's success stemmed from his ability to integrate diverse units into a cohesive system, exploiting enemy vulnerabilities through speed, coordination, and psychological disruption. His most frequently used tactics, employed consistently across major battles, prioritized combined-arms and flanking:

1. **Hammer and Anvil**: Used in all major battles - augamela (331 BC), Hydaspes (326 BC), Issus (333 BC), Granicus (334 BC), Jaxartes (329 BC), Persian Gates (330 BC), Arigaeum (327 BC), Chaeronea (338 BC), Thebes (335 BC) - it pinned enemies with the phalanx while cavalry struck flanks or rear.
2. **Flanking and Oblique Attacks**: Executed at Issus (333 BC), Hydaspes (326 BC), Jaxartes River

(329 BC), Granicus (334 BC), Persian Gates (330 BC), Arigaeum (327 BC), Chaeronea (338 BC), Sagalassos (333 BC), and Thebes (335 BC), these unbalanced enemy lines.

3. **Combined-Arms Coordination**: Integrated phalanx, cavalry, hypaspists, and light infantry for synergistic effects, seen at Gaugamela (331 BC), Hydaspes (326 BC), Issus (333 BC), Granicus (334 BC), Jaxartes (329 BC), Persian Gates (330 BC), Arigaeum (327 BC), Chaeronea (338 BC), Thebes (335 BC), and Sagalassos (333 BC).
4. **Commander Targeting**: Employed at Issus (333 BC), Gaugamela (331 BC), Hydaspes (326 BC), Persian Gates (330 BC), Granicus (334 BC), Jaxartes (329 BC), Arigaeum (327 BC), and Chaeronea (338 BC), it triggered enemy collapse by attacking leaders.
5. **Siege Innovation**: Used at Tyre (332 BC), Sogdian Rock (327 BC), Halicarnassus (334 BC), Gaza (332 BC), Multan (325 BC), Aornos (326 BC), Cyropolis (329 BC), and Sangala (326 BC), it overcame fortified defenses.

Tactical and Strategic Lessons:

- Employ Hammer and Anvil for Decisive Strikes
- Execute Flanking and Oblique Attacks for Local Superiority
- Integrate Combined-arms for Synergy
- Master Logistical Planning
- Target Enemy Commanders to Trigger Collapse
- Innovate in Siege Warfare to Overcome Defenses
- Plan for Succession

13
Oda Nobunaga (1534-1582)

Oda Nobunaga, a Japanese daimyo during the Sengoku period, favored aggressive offensive strategies backed by meticulous planning to win battles. While he employed several signature tactics, his basic winning approach was using firearms – a relatively new military technology in Japan. He integrated Portuguese arquebuses into samurai warfare, developing volley fire suited to Japanese fighting, which thereby revolutionized Japanese warfare. His battles paved the way for Japan's later unification under Toyotomi Hideyoshi and Tokugawa Ieyasu.

Nobunaga's signature tactic was **volley fire** with arquebus firearms, arranged in rotating lines often behind wooden barricades. This innovative use of arquebus matchlock firearms gave him a significant edge over rivals and revolutionized Japanese warfare.

At the Battle of Nagashino (1575), his ~3,000 gunners were arranged in three lines to form a rotating volley system of continuous fire that killed ~10,000-12,000 men in Takeda Katsuyori's infantry and cavalry forces, previously considered the most formidable fighting force in Japan.

Nobunaga's basic strategy was to use **surprise attacks** and **rapid operations**, which would destabilize slower enemies. At Okehazama (1560), Nobunaga's ~2,000 troops defeated Imagawa Yoshimoto's ~25,000 through a surprise attack aided by a storm. Swift strikes at Anegawa (1570) overwhelmed Azai and Asakura forces.

Nobunaga's swift, decisive attacks, combined with deception and divide-and-conquer strategies, enabled him to repeatedly defeat larger, established opponents. He used the velocity and shock of rapid strikes to prevent drawn-out warfare, and his swift strikes destabilized slower daimyo rivals because he moved faster than they could react.

Nobunaga also masterfully played rivals against each other, dividing them through political maneuvering (**divide-and-conquer diplomacy**) to prevent unified opposition against him. He regularly outmaneuvered nascent coalitions by striking swiftly at their weakest points before alliances could coalesce, and forged strategic alliances to isolate enemies by bringing different factions into his fold. At Okehazama (1560), he attacked Imagawa before allies could reinforce him, while an alliance with Tokugawa Ieyasu isolated enemies like Takeda and Uesugi.

Unlike his tradition-bound rivals, Nobunaga **integrated different types of military forces** into a cohesive system, adapting to battlefield conditions.

Arquebusiers delivered ranged firepower, *ashigaru* (peasant foot soldiers) used spears for defense against cavalry, traditional samurai executed shock attacks and finishing engagements, and cavalry was used in flanking maneuvers and pursuit. This versatile combined-arms approach allowed him to adapt to different battlefield conditions and opponents. At Nagashino, this system crushed Takeda forces – arquebusiers broke Takeda cavalry, *ashigaru* held lines, and samurai-cavalry pursued. At Anegawa (1570), the cavalry flanked, arquebusiers fired, and samurai engaged the enemy.

Nobunaga also promoted officers based on ability rather than birth, an example being General Toyotomi Hideyoshi who rose from peasant origins. This practice attracted capable commanders who knew they could advance through merit. He basically abandoned traditions for pragmatic strategies that worked.

His **psychological warfare** tactics included brutal intimidation, ruthlessly massacring defeated foes and rebellious allies to terrify enemies and deter resistance. His ruthlessness signaled resolve, compelling capitulation. He waged war against powerful Buddhist monasteries that had become states within states with their own territories and armies of warrior monks. The burning of Mt. Hiei (1571) killed thousands of warrior monks and others, showcasing his ruthless intimidation and willingness to break long-standing cultural and religious norms, which terrified other rebel groups. His fearsome reputation often led enemies to surrender without fighting. However, Nobunaga failed to consolidate his power through political alliances, which alienated many powerful factions.

Nobunaga was also known to employ **economic warfare** against enemies, disrupting their resources through blockades, trade restrictions, and destruction of agricultural and commercial assets. This tactic of starving opponents weakened enemies like the Ikko-ikki and rival daimyo, who then faced difficulties in sustaining their armies, forcing a surrender or collapse.

He also revolutionized **castle warfare** by building innovative fortresses that served as political and defensive strongholds. Serving as logistical hubs they acted as forward operation bases. One of his primary fortifications, Azuchi castle, marked a departure from traditional earthen fortresses in design and symbolized Nobunaga's power.

Oda Nobunaga's success during the Sengoku period stemmed from his revolutionary integration of firearms, combined-arms tactics, surprise attacks, and intimidation supported by meticulous planning and political maneuvering. Nobunaga integrated infantry, cavalry and firearms into a cohesive system where his infantry would hold the line, cavalry would attack the flanks, and arquebusiers would provide continuous firepower – creating a military advantage that transformed Japanese battle tactics. His tactics and strategies, used consistently to win battles, exploited Japan's fragmented political situation and technological advancements. However, his poor diplomacy and over-reliance on fear and terror created enemies among Japan's clergy and traditionalist nobles so that while he was militarily proficient, his methods were politically unsustainable.

Analysis of Nobunaga's Tactics:

Nobunaga's success stemmed from his integration of new technologies (arquebuses), combined-arms coordination, and ruthless pragmatism, exploiting Japan's fragmented political landscape. His most frequently used tactics, employed consistently across battles, prioritized firepower and speed:

1. **Volley Fire with Arquebuses**: Used at Nagashino (1575), Anegawa (1570), Ishiyama Hongan-ji (1570–1580), Tennōji (1582), Tedorigawa (1577), Mikatagahara (1572), Muraki (1554), Nagaragawa (1556), and Osaka (1578), it delivered continuous firepower to break enemy formations.
2. **Surprise Attacks and Rapid Operations**: Executed at Okehazama (1560), Anegawa (1570), Kanegasaki (1570), Noda-Fukushima (1570), Mikatagahara (1572), Odani Castle (1573), Muraki (1554), Nagaragawa (1556), and Tedorigawa (1577), these destabilized larger, slower foes.
3. **Combined-Arms Coordination**: Integrated arquebusiers, *ashigaru*, samurai, and cavalry at Nagashino (1575), Anegawa (1570), Kanegasaki (1570), Noda-Fukushima (1570), Odani Castle (1573), Tedorigawa (1577), Muraki (1554), Nagaragawa (1556), and Ishiyama Hongan-ji (1570–1580) for cohesive assaults.
4. **Divide-and-Conquer Diplomacy**: Employed for Okehazama (1560), Anegawa (1570), Odani Castle (1573), Kanegasaki (1570), Noda-

Fukushima (1570), Ishiyama Hongan-ji (1570–1580), Inabayama (1567), Nagashima (1574), it isolated enemies.

5. **Psychological Intimidation**: Used at Mt. Hiei (1571), Nagashima (1574), Ishiyama Hongan-ji (1570–1580), Odani Castle (1573), Noda-Fukushima (1570), Anegawa (1570), Inabayama (1567) and Okehazama (1560), against defeated foes, it deterred resistance.
6. **Economic Warfare**: Seen at Ishiyama Hongan-ji (1570–1580), Nagashima (1574), Inabayama (1567), Okehazama (1560), Odani Castle (1573), Kanegasaki (1570), Osaka (1578), and Nagashima (1574), this weakened enemy resources and forced surrender.
7. **Innovative Fortifications**: Azuchi Castle (1576–1579), Inabayama (1567), Odani Castle (1573), Ishiyama Hongan-ji (1570–1580), Nagashima (1574), Gifu Castle (1570), Kanki Castle (1578), and Iwamura (1572), supported campaigns and projected his power.

Tactical and Strategic Lessons:

- Adopt Emerging Technologies
- Use Divide-and-Conquer Diplomacy to Isolate Enemies (Exploit Political Fragmentation)
- Disrupt Enemy Logistics
- Execute Surprise Attacks for Rapid Victories
- Integrate Combined-arms for Versatility
- Leverage Psychological Intimidation to Deter Resistance
- Promote Merit-Based Leadership for Capability

14
Mustafa Kemal Atatürk
(1881-1938)

Mustafa Kemal Atatürk, the founder of modern Turkey, was a Turkish field marshal who excelled at defensive warfare during the Gallipoli Campaign and Turkish War of Independence (1919–1922), winning battles like Sakarya and Dumlupınar. By blending terrain-based defense with decisive counterattacks against overextended Greek and Allied forces, his tactics, bolstered by nationalist fervor, secured Turkish independence.

Atatürk's signature tactic was **elastic defense** where he lured enemies into overextended positions (with deception as a supportive element), absorbed their assaults with multiple defensive lines (layered defenses), and then launched rapid counterattacks. He frequently created false impressions of strength or weakness to mislead enemies, launched diversionary attacks, and employed false retreats to lure Greek and Allied forces into traps. He would appear to leave positions open or withdraw, only to launch a counteroffensive with reinforcements at decisive moments. At Sakarya (1921), he ceded ground to the Greeks in order to stretch the Greek supply lines, then counterattacked with fresh troops to repel the Greeks, forcing their retreat. At Dumlupınar (1922), this tactic expelled Greek forces, securing him a victory.

This elastic basic battle strategy, exemplified at Sakarya (1921), would progress through the following sequential steps before finally culminating in a counterattack:

1. Luring the enemy to advance into difficult terrain where they would overextend (ceding ground to stretch Greek lines)
2. Absorbing their attacks with strong defensive positions, allowing enemy forces to exhaust themselves in initial assaults
3. Identifying the moment when attacking forces were most vulnerable
4. Launching decisive counterattacks with fresh troops at the enemy's weakest points

A defensive commander, Atatürk **exploited terrain** by choosing battlegrounds that could *naturally* neutralize enemy advantages, using high ground and natural obstacles to afford defensive protection. At Gallipoli (1915), positions on Conkbayırı (Chunuk Bair) repelled Anzac assaults. At Sakarya, the rugged Anatolian terrain disrupted the Greeks' logistics. When Turkish forces were outnumbered, he used the terrain to funnel enemies into narrow kill zones to ensure Turkish dominance despite being outgunned.

Atatürk maximized the usage of terrain for defense, but his elastic approach also emphasized **dynamic counterattacks**. He concentrated his forces to attack vital enemy points at opportune moments to achieve local superiority wherever possible. Atatürk maintained sufficient **reserves** for concentrated counterattacks at any enemy weak points. For instance, at Sakarya, fresh troops struck the overextended Greeks. At Gallipoli, counterattacks at Conkbayırı repelled enemy breakthroughs. The precise timing of attacks was chosen to exploit enemy exhaustion, achieve local superiority and amplify the Turkish impact despite Atatürk's limited resources.

Atatürk commonly used **deception**, including feigned withdrawals, to support his elastic defense, luring enemies into vulnerable positions before counterattacking. At Gallipoli, feigned withdrawals at Suvla Bay lured Anzacs into ambushes. In the Great Offensive (1922), diversions concealed the assaults at Dumlupınar, including Afyon.

Another tactic that Atatürk employed was **guerrilla and irregular warfare**. He used small, mobile units to harass enemy supply lines, disrupt enemy communications, and weaken their morale. These tactics, often led by local militias or irregular forces, complemented his conventional operations, and were critical to stretching enemy resources and maintaining pressure.

Turkish forces, under Atatürk's leadership, also used **scorched earth tactics** to disrupt Greek supply lines and weaken their forces, as seen in 1921-1922. Atatürk's forces avoided widespread destruction to maintain civilian support but he employed limited scorched earth tactics to deny resources to the Greeks at Sakarya (1921). Burning crops and villages weakened Greek forces prior to Turkish counterattacks to recapture lost territory and force the enemy to retreat.

Atatürk's tactics – terrain exploitation (using high ground and natural obstacles), elastic defenses, deception, skillful management of reserves, counterattacks, and scorched earth – overcame superior enemies. He fought his battles with both tactics and national purpose to win Turkish independence against overwhelming odds.

A portion of his success must be attributed to nationalist fervor alongside his tactics and military skill because – similar to other revolutionary wars – the Turkish resistance depended on widespread support rather than just tactical brilliance. Like Wellington, Atatürk maximized the use of terrain, but his dynamic counteroffensives defined his legacy as a defensive tactician.

Analysis of Atatürk's Tactics:

Atatürk's success stemmed from his mastery of defensive warfare, blending terrain-based strategies with dynamic counterattacks to exploit enemy weaknesses. His most frequently used tactics, employed consistently across battles, prioritized elastic defense and deception:

1. **Elastic Defense with Counterattacks**: Used at Sakarya (1921), Dumlupınar (1922), Gallipoli–Conkbayırı (1915), Anafartalar (1915), Kütahya-Eskişehir (1921), First İnönü (1921), Tobruk (1911), Kovalca (1921), and Second İnönü (1921), it lured enemies into overextended positions before counterattacking.
2. **Terrain Exploitation**: Leveraged at Gallipoli–Conkbayırı (1915), Dumlupınar (1922), Anafartalar (1915), First İnönü (1921), Second İnönü (1921), Kütahya-Eskişehir (1921), Sakarya (1921), Afyonkarahisar (1922), and Chunuk Bair (1915), it neutralized enemy advantages.
3. **Deception (Feigned Withdrawals and Diversions)**: Employed at Gallipoli–Suvla Bay (1915), Dumlupınar (1922), Sakarya (1921), Anafartalar (1915), First İnönü (1921), Second İnönü (1921), Kovalca (1921), and Kütahya-Eskişehir (1921), it misled enemies into traps.
4. **Guerilla Warfare**: Seen at Sakarya (1921), Dumlupınar (1922), Great Offensive (1922), Kütahya-Eskişehir (1921), Afyonkarahisar (1922), First İnönü (1921), Kovalca (1921), and the Second İnönü (1921).
5. **Scorched Earth Tactics**: Used selectively at Sakarya (1921), Dumlupınar (1922), Kütahya-Eskişehir (1921), Great Offensive (1922), Afyonkarahisar (1922), First İnönü (1921), and Second İnönü (1921), it disrupted enemy logistics.
6. **Reserve Concentration for Local Superiority**: Executed at Sakarya (1921), Gallipoli–Conkbayırı (1915), Dumlupınar (1922), Anafartalar (1915), First İnönü (1921), Second İnönü (1921),

Kütahya-Eskişehir (1921), and Afyonkarahisar (1922), it amplified counterattack impact.

Tactical and Strategic Lessons:

- Employ Elastic Defense with Decisive Counterattacks
- Exploit Terrain to Neutralize Enemy Advantages
- Exploit Enemy Overextension
- Use Deception to Lure Enemies into Traps
- Integrate Irregular Warfare
- Apply Limited Scorched Earth to Disrupt Logistics
- Concentrate Reserves for Local Superiority
- Harness Nationalist Fervor to Bolster Morale

15
Ferdinand Foch (1851-1929)

Ferdinand Foch served as a French military commander during World War I, eventually becoming Supreme Allied Commander during the latter part of the conflict. His tactical approach evolved dramatically over the course of the war, starting from a disastrously aggressive "offensive at all costs" doctrine in 1914 to a sophisticated, combined-arms approach of methodical battles targeting limited objectives that ultimately contributed significantly to Allied victory in 1918.

During the war, Foch **coordinated the efforts of multiple national armies** – French, British, American, Italian, and others – while balancing competing national interests and military cultures. He developed a common strategic vision across multiple armies, creating a synchronized, cohesive front against enemy forces. His approach to unified operations later became a blueprint for the Allied military operations of World War II.

Beyond coordinating the forces of different nations, Foch integrated new technologies – tanks, aircraft, and artillery – with traditional infantry to break trench stalemates. At Amiens (1918), tanks and timed barrages achieved a 7-12 mile advance. In Meuse-Argonne (1918), combined-arms and attrition breached German lines. This combined-arms approach of new technologies and tactics effectively overcame entrenched defenses that previously resulted in deadlock.

Foch's early strategic vision centered on aggressive offensives as the primary path to victory. He rigidly adhered to the French military doctrine of "*offensive à outrance*" (attack at all costs), which held that determined infantry could overcome defensive firepower through aggressive action. However, this approach lead to disastrous early battles. In line with this philosophy, he initially championed mass frontal assaults of sheer manpower and firepower to grind the enemy down. This approach, based on the belief that morale and willpower could overcome firepower, proved catastrophic at the Battle of the Frontiers in 1914, where French forces suffered heavy losses to German machine guns, artillery and maneuver.

While Foch initially embraced the French doctrine of "*offensive à outrance*," which contributed to early disasters, he eventually adapted after setbacks. Following early failures, Foch reassessed his pre-war doctrines to recognize that uncoordinated attacks without preparation and proper artillery support were ineffective. He improved artillery coordination, using timed barrages to support advances, although breakthroughs remained elusive.

While maintaining his focus on offensive action, he shifted from pure offense to the "**methodical battle**," which became his signature tactic. He started seeking step-by-step advances – attacks for limited objective to erode German strength – rather than grand breakthroughs. In other words, instead of seeking a single decisive battle, the new Foch battle method that emerged in 1918 began going after limited objectives in a sequential fashion. The Foch methodical battle approach featured:

- Limited-objective attacks instead of breakthrough attempts
- Sequential "hammer blows" along different sectors of the front
- "Bite and hold" tactics aimed at gradually eroding German strength

Foch also employed a "**hammering strategy**" (*martelage*) that involved sequential attacks along multiple sectors of the front to create a cumulative pressure that would incrementally destabilize the enemy. These sequential hammer blows across a broad front were designed to exhaust the enemy's reserves and convert small gains into larger strategic victories. The hammering strategy reflected the realization that the war would be won through sustained pressure rather than speed. Determining that victory would require this sustained pressure rather than a single massive attack, in the Hundred Days Offensive, sequential artillery assaults (as part of a combined-arms offensive) at Ypres, Amiens, and

Meuse-Argonne protected advancing troops and collapsed German defenses.

While implementing **flexible defenses** – multiple lines of defense to absorb enemy advances and cause them heavy casualties while allowing Allied forces to retreat to stronger positions – Foch always viewed defense as only a temporary measure, a prelude to counteroffensive attacks at the enemy's weakest points. Therefore, during the German Spring Offensive (March-June 1918) Foch coordinated flexible defenses, holding key positions before launching counterattacks. He preferred launching counteroffensives to regain the initiative after the enemy had overextended themselves, as demonstrated at the Second Battle of the Marne (July 1918).

Foch skillfully **managed reserves**, ensuring that fresh troops were always available to reinforce critical battlefield positions or exploit breakthroughs. He strategically held back elite units, typically American or French troops, for well-timed counterattacks. At the Second Battle of the Marne (1918), reserves halted German advances, enabling counteroffensives while during the Hundred Days Offensive (1918) that ultimately led to the end of WWI, fresh troops sustained Allied momentum while the exhausted Germans were unable to replenish their forces.

Ultimately, Foch's **attritional strategy** exhausted German reserves through relentless offensives that maintained maximum pressure on all fronts while depleting German resources, shattering Germany's ability and will to fight. This helped force the Central Powers into an armistice. His battlefield success evolved from a rigid Napoleonic offensive doctrine that dominated pre-war French military thinking to a system of *sophisticated, synchronized, multi-pronged, combined-arms offensive actions across different armies*. In time, Foch's advanced tactics – methodical battle, hammer blows, combined-arms, reserve management, and coordination – broke trench warfare stalemates and shattered German resistance. In 1918, his synchronized assaults overwhelmed the Germans.

Foch's **diplomatic coordination** unified French, British, and American armies into an effective fighting force, aligning their competing interests and efforts into a cohesive strategy. Some argue that Foch was more of a diplomatic general than a tactical genius because his diplomatic finesse that united Allied armies seemed more impressive than his battlefield command, but his tactical vision and operational innovations (e.g., methodical battle) drove battlefield success. His strategic pressure and armistice negotiations with Germany hastened the Central Powers' collapse.

Analysis of Foch's Tactics:

Foch's success stemmed from his ability to adapt pre-war doctrines, coordinate multinational forces, and integrate new technologies to break trench warfare stalemates. His most frequently used tactics, employed consistently in 1918, prioritized methodical, multi-pronged offensives:

1. **Methodical Battle with Limited Objectives**:
   Used in the Hundred Days Offensive (1918),

Amiens (1918), Meuse-Argonne (1918), St. Mihiel (1918), Grand Couronné (1914), Soissons (1918), Montdidier-Noyon (1918), Aisne-Marne (1918), and Flanders (1918), it eroded German strength through step-by-step advances.
2. **Sequential Hammer Blows (*Martelage*)**: Executed across the Hundred Days Offensive (1918), Amiens (1918), Meuse-Argonne (1918), Ypres (1918), Soissons (1918), St. Mihiel (1918), Montdidier-Noyon (1918), Lys (1918), and Marne (1918), these pressured multiple front sectors.
3. **Combined-Arms Integration**: Leveraged tanks, aircraft, and artillery at Amiens (1918), Meuse-Argonne (1918), Soissons (1918), St. Mihiel (1918), Marne (1918), Ypres (1918), Lys (1918), Aisne-Marne (1918), and Vittorio Veneto (1918), to breach defenses.
4. **Flexible Defense with Counterattacks**: Employed at the Second Battle of the Marne (1918), Spring Offensive (1918), Amiens (1918), Soissons (1918), Ypres (1918), St. Mihiel (1918), Lys (1918), and Grand Couronné (1914), it absorbed enemy assaults before counteroffensives.
5. **Reserve Management**: Used in Marne (1918), Spring Offensive (1918), Amiens (1918), Soissons (1918), Ypres (1918), St. Mihiel (1918), Montdidier-Noyon (1918), Aisne-Marne (1918), and Grand Couronné (1914), it sustained momentum with fresh troops.

Tactical and Strategic Lessons:

- Adapt Doctrine to Realities

- Exploit Technological Advances
- Employ a Methodical Battle Approach for Steady Gains
- Execute Sequential Hammer Blows Across Fronts
- Integrate Combined-arms to Break Stalemates
- Use Flexible Defense as a Prelude to Counterattacks
- Manage Reserves for Sustained Momentum
- Coordinate Multinational Forces for Unity
- Foster Diplomatic Cohesion

16
Douglas Haig (1861-1928)

Field Marshal Douglas Haig, commander of the British Expeditionary Force (BEF) in World War I, championed a doctrine of attrition warfare that defined the Western front. He advocated continuous, aggressive action across the entire front to break enemy lines, which resulted in staggering British casualties. Nonetheless, his strategies at the Somme, Passchendaele, and the Hundred Days Offensive wore down German forces, securing Allied victory.

Haig initially believed that breaking the deadlock of trench warfare required massive, sustained infantry offensives. His conviction that overwhelming force (large-scale infantry attacks) would eventually rupture enemy lines drove him to accept enormous casualties from frontal assaults, even when the attacks were unsuccessful, because he assumed that the continuous pressure would eventually lead to a breakthrough by wearing down German forces until they could no longer fight.

Haig relied on preliminary **massive artillery bombardments** as the cornerstone of his approach. He ordered days – sometimes weeks – of steady heavy shelling intended to soften enemy positions. This aimed to destroy barbed wire, neutralize machine guns, and decimate enemy ranks before sending in British infantry. However, the barrages often failed to destroy German defenses, leaving British infantry as easy targets for German gunners when they advanced in formations. The Somme's week-long barrage failed to destroy German lines, costing 57,470 British casualties (19,240 killed) on the first day of the infantry offensive (July 1, 1916). Although a tactical failure, the campaign aimed to wear down German forces and relieve Verdun so it achieved some partial strategic success. By 1918, precise barrages at Amiens supported successful assaults, reflecting improved targeting and coordination.

Even when there were no immediate gains from the previous infantry failures and the staggering casualties demoralized his own troops, Haig initially remained steadfast in his tactics, arguing that he was eventually wearing down German fighting power. He was willing to accept heavy British casualties in order to inflict proportional or greater losses on German forces. He truly believed that Germany could be defeated by gradually eroding its manpower and resources, conceiving of battles *as part of this wearing-down process* rather than as decisive encounters. His basic attritional strategy involved:

1. Massive artillery bombardments targeting enemy defenses
2. Large-scale infantry assault waves to break though enemy lines
3. The commitment of reserves to exploit any gains

Haig was initially cautious about unproven technologies like tanks but embraced innovations after trials, as seen at Cambrai (1917) where he began deploying tanks en masse for the first time to break through entrenched lines. There, supported by artillery, 378 tanks marked the first large-scale tank offensive. He gradually developed a more sophisticated **combined-arms approach** that integrated tanks, aircraft and infantry in coordinated battlefield attacks. At Amiens, tanks, aircraft and creeping barrages successfully led to a 7-12 mile advance in the front.

Haig also shifted from a purely attritional approach to a more measured "**bite and hold**" **strategy** that focused on achieving limited objectives with precise attacks, rather than breakthroughs. The "bite and hold" strategy involved attacking in stages ("biting") to secure small, achievable gains followed by their consolidation before advancing further ("holding"). During the Hundred Days Offensive (1918), battles like Amiens (1918) and Canal du Nord (1918) captured their objectives, and the gains were then consolidated to prevent German recovery. At Passchendaele (1917), later phases of the battle used "bite and hold" to secure ridges despite muddy conditions.

This more refined approach to battles ensured sustainable progress. It essentially meant a change from attritional warfare to limited, well-planned advances to seize key objectives, consolidate gains, and repel German counterattacks. Haig would launch precise artillery bombardments targeting enemy defenses followed by limited infantry assaults to secure specific objectives, rapidly consolidate the gained positions, and employ reserves to exploit breakthroughs.

Haig adopted a **"learning curve" approach of iterative improvements** to refine his battle tactics through testing and analysis, a practice shared with other commanders. He would test new tactics in small engagements, analyze their results through rigorous staff work, and then implement successful innovations army-wide. Effective solutions were scaled up for major offensives; he modernized gradually under fire. This approach of gradual improvements was evident in the expansion of the British Expeditionary Force (BEF) from six divisions in 1914 to over sixty divisions (including Dominion units) by 1918. This iterative process refined BEF effectiveness over time.

Additionally, Haig oversaw the BEF's **massive logistical expansion** to sustain operations, which included the construction of extensive railway networks behind British lines and the creation of enormous supply dumps. This logistical planning detailed the provision of ammunition supply, food, and other necessities, alongside the development of medical evacuation chains to process mass casualties. He also oversaw reserve management to sustain operations and exploit gains, ensuring fresh troops were available for critical moments.

Haig's tactics – bombardments, combined-arms, bite and hold, and logistics – evolved from the Somme's costly failures to 1918's successes. He initially believed attrition (massive artillery bombardments and infantry assaults) would wear down the enemy over time, so he did not seek quick victories nor rapidly adapt to new technologies at first. However, his tactics evolved significantly over the course of the war. Learning from the mistakes that led to bloody slaughter, his initial attritional warfare strategy gradually evolved into a more sophisticated and effective combined-arms approach, complemented by a "bite and hold" strategy that produced meaningful gains with fewer casualties.

Analysis of Haig's Tactics:

Haig's success stemmed from his persistence in attritional warfare and eventual adaptation to combined-arms and limited-objective tactics, leveraging Britain's industrial and logistical strengths. His most frequently used tactics, employed consistently across major battles, evolved over time:

1. **Attrition Warfare with Massive Bombardments**: Used at the Somme (1916), Passchendaele (1917), Arras (1917), Loos (1915), Neuve Chapelle (1915), Messines (1917), Le Transloy (1916), Vimy Ridge (1917) and Menin Road (1917), it aimed to erode German manpower and exhaust their resources.

2. **Bite and Hold Strategy**: Adopted in the Hundred Days Offensive (1918), Amiens (1918), Passchendaele–Menin Road (1917), Messines (1917), Polygon Wood (1917), Broodseinde (1917), Havrincourt (1918), and Canal du Nord (1918), it secured sustainable gains.
3. **Combined-Arms Integration**: Implemented at Cambrai (1917), Amiens (1918), Hundred Days (1918), Canal du Nord (1918), St. Quentin Canal (1918), Messines (1917), Menin Road (1917), Vimy Ridge (1917), Havrincourt (1918), and Menin Road (1917), it broke trench stalemates with tanks, aircraft, and infantry.
4. **Learning Curve Iterative Improvements**: Applied throughout the war, it refined tactics through testing and analysis. Seen at Cambrai (1917), Messines (1917), Menin Road (1917), Amiens (1918), Neuve Chapelle (1915), Arras (1917), Vimy Ridge (1917), Le Transloy (1916)and St. Quentin Canal (1918).
5. **Logistical Expansion and Reserve Management**: Sustained operations at the Somme (1916), Passchendaele (1917), Hundred Days (1918), Amiens (1918), Cambrai (1917), Arras (1917), Vimy Ridge (1917), Havrincourt (1918), and St. Quentin Canal (1918).

Tactical and Strategic Lessons:

- Employ Attrition Warfare to Deplete Enemy Resources
- Adapt to Technological Evolution
- Adopt Bite and Hold for Sustainable Gains

- Integrate Combined-arms to Breach Defenses
- Implement Iterative Improvements via Learning Curve
- Develop Robust Logistics to Sustain Operations
- Manage Reserves to Exploit Breakthroughs

17
Augustus Caesar
(63 BC-14 AD)

Augustus Caesar (born Octavian), Rome's first Emperor, was renowned for his political acumen, use of propaganda, and administrative skill in consolidating power – not military prowess. For instance, he formed the Second Triumvirate with Mark Antony and Lepidus to defeat Caesar's assassins, and used diplomacy to consolidate his power and avoid direct military conflict after the defeat of Brutus and Cassius at the Battle of Philippi.

Augustus rarely personally commanded armies in battle and was more of a strategic military-political leader than a battlefield tactician, who delegated battlefield command to generals while directing campaigns and diplomacy. His great strength was recognizing and empowering exceptional military commanders such as Marcus Vipsanius Agrippa who designed and executed most military operations, Tiberius (his eventual successor), and Drusus. Agrippa's victories at Naulochus (36 BC) defeated Sextus Pompey, and at Actium (31 BC) secured naval dominance. Tiberius subdued the Cantabrian tribes (26–19 BC) through attrition. Augustus didn't win battles but won control of the men who could fight them, succeeding militarily by delegating tactical command to the right people.

Augustus avoided direct military confrontation whenever possible, preferring diplomacy and strategic alliances. He would try to isolate enemies diplomatically before fighting. When choosing to engage in battle, he made sure he had strongly favorable odds. He maintained overall strategy while delegating combat leadership to capable commanders, particularly Agrippa.

Augustus's signature **"divide and conquer"** battle strategy combined diplomacy, economic pressure, and targeted military force to weaken and defeat enemies. You could also call this "diplomatic isolation" because Augustus would attempt to fracture enemies by neutralizing their allies. This involved a three-phase pathway to success:

1. Politically isolate enemies by turning their allies neutral
2. Block supply lines for economic strangulation
3. Push for a decisive winning battle

This approach minimized actual combat: destabilize enemies politically through alliances and propaganda, force wars of attrition to deplete their resources, and then strike decisively against the weakened foes. At Actium, alliances with Antony's former allies isolated Mark Antony, ensuring victory. To subdue the Illyrian and Alpine tribes, he exploited internal rivalries to neutralize allies by fracturing tribal unity before striking. This exemplified his tactic of divide and conquer. His approach combined political manipulation (alliances and coalitions) with targeted military force to pacify the region.

Rather than annexing every territory through military conquest, Augustus negotiated alliances with local rulers like Herod the Great in Judea, offering them Roman protection and support in exchange for loyalty. In 25 BC, the kings of Mauretania became Roman vassals for the Roman control of North Africa. By setting up **client-king buffer states**, he could maintain Roman control of a region without direct Roman occupation. He prioritized natural boundaries like the Rhine and Danube rivers, a policy reinforced after the catastrophic loss of three legions to Germanic tribes at Teutoburg Forest (9 AD). Diplomatic negotiations with powers like Parthia reduced the need for costly military campaigns.

Augustus often employed strategic patience in his conflicts, and showed a willingness to engage in protracted campaigns to wear down enemies. His campaigns against Cantabrian tribes in Hispania (26-19 BC) exemplified this grinding approach. One of his tactics was economic attrition, using blockages and sieges to starve enemies. In the Perusine War (41–40 BC), he starved the city's defenders into surrendering; his strategic patience like this ensured favorable odds for winning with minimal losses. Another example is at Actium (31 BC) where Agrippa's blockade weakened Antony's fleet, leading to a decisive naval victory that secured Augustus's unchallenged rule of Rome.

Augustus built upon Julius Caesar's battlefield tactics to achieve greater flexibility: spacing soldiers apart, and standardizing flexible *maniple* units (a subdivision of a Roman legion) for tactical adaptability. He extensively used field fortifications to defend positions, secure flanks, and besiege enemy strongholds. He systematized frontier defenses by building permanent military camps along frontiers and creating client buffer kingdoms, as stated, to buffer potential threats. He would Romanize new provinces through military roads and veteran settlements, thus turning war into a permanent civil order.

Augustus also implemented sweeping military reforms that standardized the Roman army, establishing a standing army of ~28 legions (~150,000–200,000 men), and creating the Praetorian Guard and auxiliary units. He systematized infantry pay, veteran settlements, and frontier infrastructure. These military reforms were foundational to the empire's ability to manage multiple fronts.

Augustus's strategies – diplomatic isolation, attrition, delegation to capable commanders, client states and buffer kingdoms, and flexible tactics – built a durable empire. Though he lacked the personal battlefield legacy of figures like Caesar or Hannibal, Augustus's military strategies ensured Rome's dominance for centuries. He systematized Rome's approach to conquest and defense by building a military machine that functioned whether he was present or not. This allowed the Roman Empire to fight multiple conflicts simultaneously across distant fronts.

After securing power, Augustus faced the challenge of maintaining peace and stability across vast territories. He became an expert at winning the hearts and minds of the people and established the Pax Romana through diplomacy and military reforms that ushered in an unprecedented era of peace, which fostered economic and cultural prosperity throughout the Empire.

Analysis of Augustus's Strategies:

Augustus's success stemmed from his strategic acumen, delegating battlefield command to capable generals while orchestrating diplomacy, economic pressure, and targeted military force. His most frequently used strategies, employed consistently across campaigns, prioritized minimal combat and long-term stability:

1. **Divide and Conquer (Diplomatic Isolation)**: Used at Actium (31 BC), Illyria (35–33 BC), Alps (25–15 BC), Philippi (42 BC), Naulochus (36 BC), Pannonian Revolt (6–9 AD), Sicilian Campaign (36 BC), Dalmatian Campaign (12–9 BC), and in the Cantabrian Wars (26–19 BC), it fractured enemies through alliances and propaganda.
2. **Economic Attrition and Blockades**: Employed in the Perusine War (41–40 BC), Actium (31 BC), Cantabrian Wars (26–19 BC), Naulochus (36 BC), Illyria (35–33 BC), Alps (25–15 BC), Sicilian

Campaign (36 BC), and Pannonian Revolt (6–9 AD), it starved enemies into submission.
3. **Delegation to Capable Commanders**: Seen at Naulochus (36 BC), Actium (31 BC), Cantabrian Wars (26–19 BC), Illyria (35–33 BC), Alps (25–15 BC), Pannonian Revolt (6–9 AD), Dalmatian Campaign (12–9 BC), and the Raetian Campaign (15 BC), ensuring tactical excellence.
4. **Client States and Buffer Kingdoms**: Established with Mauretania (25 BC), Judea (31 BC–14 AD), Thrace (20 BC–14 AD), Armenia (20 BC), Noricum (16 BC), and Pontus/Cappadocia (20 BC–14 AD), and German Frontier Campaigns (12 BC–9 AD), they reduced direct military burdens.
5. **Flexible Tactics and Fortifications**: Standardized maniples and field fortifications in Cantabria and along frontiers, enhancing adaptability. Flexible tactics seen in the Cantabrian Wars (26–19 BC), Illyria (35–33 BC), Alps (25–15 BC), Pannonian Revolt (6–9 AD), Teutoburg Forest (9 AD), Naulochus (36 BC), Dalmatian Campaign (12–9 BC), German Frontier Campaigns (12 BC–9 AD), and the Raetian Campaign (15 BC).

Tactical and Strategic Lessons:

- Employ Divide and Conquer for Diplomatic Isolation
- Use Economic Attrition to Starve Enemies
- Delegate to Capable Commanders for Tactical Success
- Establish Client States for Strategic Control

- Integrate Propaganda for Legitimacy
- Adopt Flexible Tactics and Fortifications
- Secure Natural Frontiers
- Standardize Military Systems
- Exercise Strategic Patience for Favorable Odds
- Balance Conquest with Good Governance
- Win Hearts and Minds for Stability

18
Final Analysis & Thoughts

If we analyze the track records of these winning generals we will find a strong bias towards aggressive, offensive warfare. This clearly reveals the most successful approach for winning wars. Only two generals out of our fifteen – the Duke of Wellington and Mustafa Kemal Atatürk – can be classified as primarily defensive generals, and both became masters of terrain-based defenses.

Because Wellington and Atatürk preferred the defensive warfare approach of absorbing assaults and then counterstriking at the appropriate time, they chose battlefields that forced enemies to attack on unfavorable terms. While every general knows to secure maximum terrain advantages, the principle for defense should be to incorporate natural barriers into your defensive plans, as they did, because they make your defense as strong as possible. If you are going to go defensive, choose locations that are already maximally defensive and employ their strength.

Let us also set aside the topic of psychological warfare, intelligence, military deception and luring enemies into traps, which are now well-understood topics because most commanders have studied Sun Tzu. All generals know to employ these tactics, so we need not elaborate further. This is why I did not delve deeply into the methods each general used for psychological warfare and misleading their enemy.

The same goes for the topic of diplomacy. While Augustus and Nobunaga used political isolation to help defeat foes, and other generals formed multi-national coalitions, such as Wellington, because this topic is quite specialized, we won't delve deeply into it.

As an aside, the sage Guan Tzu, who was superior to Sun Tzu concerning strategies for elevating a country economically and politically while facing military foes, focused on higher-level perspectives that a nation could employ for becoming a superpower without fighting. Guan Tzu taught how and when a country should join multi-state military alliances, whether to become their leader or just one of the pack, and used various means to help his state become supreme while raising his own lord to become the predominant leader in the land even though China already had an Emperor.

The Art of Political Power: A Translation and Commentary of The Book of Master Guan (The Guan Tzu) teaches ancient and modern methods of economic warfare, as well as the strategies that nations facing various conditions, based on their size, should use to either preserve themselves among the stronger nations, become the predominant party in a coalition, or become a superpower. Some of our top generals gained control of supply chains to economically defeat enemies, but a fuller picture of economic warfare can be found in this book as well as the steps necessary to take a state to superpower status.

SPEED & MANEUVER

Before moving on to the most common tactics that our generals used for winning battles, we must also note that the majority of winning generals at the top of the list were particularly noted for executing swift troop maneuvers. Napoleon had his Corps system, Caesar used rapid forced marches, Takeda Shingen was known for his cavalry charges, Khalid Ibn al-Walid was recognized for his rapid marches across the desert and camel corps, Frederick the Great had a reputation for quick marching, Oda Nobunaga was known for his swift attacks, and so on. Rapid movements surprise enemies, so it is the purpose for modern rapid-response units.

Obviously, the need to maneuver faster and act quicker than opponents was and is a major key to winning battles. All military commanders must strive to master swift decision-making and maneuvers. I fondly recall the movie scene in *Patton* where the general said, "I can attack with three divisions in 48 hours." He actually did turn three Third Army division 90 degrees and move them north within 72 hours to counter the German offensive in the Ardennes, a remarkable feat of speed and flexibility.

Gen. George Patton, incidentally, accumulated a .9 WAR score compared to a negative WAR score of

-1.953 for Erwin Rommel, which helps settle the argument as to who was really the better commander. This has been a contentious topic over the years because many have claimed that Rommel's reputation was exaggerated due to Nazi propaganda, and that his victories were partly due to Allied mistakes rather than his own military brilliance. The sabermetric approach provides us with some objective evidence for a comparison.

INTEGRATION OF FORCES, COMBINED ARMS & NEW TECHNOLOGY

Another observation is that the most successful generals commonly integrated different military forces into a cohesive whole despite any long-standing traditions that might go against combining different classes and races.

They also used a combined-arms approach whether it be to integrate infantry, cavalry and archers or infantry, armor, artillery, navy and air power.

Most also showed little to no hesitancy to adopt and employ the latest technologies into their battle plans. This was often a game changer for winning battles, such as Grant's use of gunboats, Foch's integration of tanks and airpower, or Oda Nobunaga's use of matchlock firearms. The general with a technological edge, or new tactics, usually had the advantage in battle.

From Alexander's phalanx-cavalry synergy to Foch's tank-infantry coordination, and from the history of various multi-state alliances (including Wellington, Foch and Haig), the lesson is that combined-arms and integrating forces maximizes impact.

MOST EFFECTIVE BATTLE TACTICS

Now, as for the historical finding of the top tactics for winning battles that we see successfully used time and again by the top generals – which is the information I most sought – this short list includes:

- Flanking maneuvers (while pinning the center)
- Encirclements (while pinning the center)
- Dividing and conquering
- Attritional campaigns
- Siege warfare
- Elastic defenses to absorb enemy attacks followed by strong counteroffensives
- Skillful management of reserves
- Diplomacy to isolate enemies from their friends, and to form coalitions against enemies
- Rigorous training of standardized drills so that troops are faster and more accurate
- Pursuit and total annihilation of the enemy rather than the goal of conquering territory

The details of these tactics have already been discussed in our fifteen biographies. These are the methods that a military student must learn, and which a commander must certainly master as a priority if they want to win wars. How did the greatest generals win wars? They used these tactics and battle techniques.

THE ROLE OF DIPLOMACY

After reviewing these histories, I noticed a final unexpected commonality. Among all who sought to become emperors or just unite countries through warfare – Napoleon Bonaparte, Julius Caesar, Takeda Shingen, Oda Nobunaga, Alexander the Great, and Augustus Caesar – the only one who fared well in the end, or fared best, was the one who most emphasized diplomacy in his strategies rather than simply fighting to win an empire through conquest. This was Augustus Caesar, who successfully laid the foundation for the Pax Romana, a period of relative peace and prosperity in the Roman Empire. The practice of fighting without a larger strategy to consolidate any gains did not end well for most.

We must train ourselves to act as Themistocles did as he focused on building Athens into a naval power that would secure its future. His short-term strategies won battles, his intermediate-term vision strengthened Athens' navy, and his long-term planning ensured the city's dominance in Greece. He thought about the long-term objective of peace and prosperity, as did Augustus Caesar.

Some of the more robust and dependable methods for achieving this are summarized within *The Art of Political Power* and *Culture, Country, City, Company, Person, Purpose, Passion World*.

If one can avoid war, that is best. If one can win without fighting, that is better than fighting. War can, in an instant, destroy what took generations to build. Whatever you decide to do must incorporate the desired trajectory and future status of your nation.

THE BLOODY COST OF TOTAL WAR

The ruthlessness of some generals in pursuing enemies or in dealing with traitors has been noted, but it is only when we come to the total war and attritional warfare generals – such as Grant, Foch and Haig – that we find a larger chorus critical of the massive bloodshed, much of which was unnecessary. These generals launched massive grinding frontal assaults meant to wear down enemies, and their casualties were enormous.

This is the cruelty of war, which destroys lives and countries. By studying the most effective tactics used to win wars, however, we enhance the possibility that able commanders may learn the appropriate lessons that will enable them to achieve battlefield success more quickly to end the carnage. It is not just the goal of winning wars that matters, but winning quickly at minimal cost, and then establishing peace and prosperity thereafter.

SYNOPSIS

In examining the careers of history's most successful generals, we have distilled centuries of battlefield experience into a concise set of proven tactics and strategies. While each military commander faced unique circumstances and opponents, their victories reveal enduring patterns that transcend time and technology.

From flanking maneuvers and encirclements to elastic defenses and attritional campaigns, these military tactics have proven effective across different eras, cultures, and technological levels. The consistency of these methods suggests fundamental truths about warfare that military students must master: speed and maneuver provide decisive advantages, combined-arms approaches multiply combat effectiveness, and the ability to concentrate force at critical points while deceiving the enemy remains paramount regardless of the weapons employed.

These lessons hold particular relevance as warfare evolves into new domains. While future conflicts may feature autonomous systems, cyber operations, scalar communications, EMF or particle beam weaponry, and space-based assets, the underlying principles of warfare will likely remain unchanged. Commanders will still need to achieve local superiority, exploit enemy weaknesses, protect their own vulnerabilities, and maintain the initiative.

The tactical frameworks developed by Napoleon, Alexander, and their peers – adapted for modern conditions – will continue to guide military thinking. Understanding how Wellington maximized terrain advantages, how Grant coordinated multiple armies across vast distances, or how Zhukov orchestrated massive operational deceptions provides essential foundations for addressing contemporary military challenges. One needs to understand the principles behind the winningmost military tactics, and then adapt those tactics for modern conflicts.

Ultimately, this study serves as both a practical guide and a cautionary tale. The generals who achieved lasting success were those who applied proven tactics while adapting to their circumstances, who balanced aggressive action with strategic patience, and who understood that military victory must serve broader political objectives. The wins must be turned into a stable national success so that the benefits are not lost.

As we face an uncertain future where traditional battlefield boundaries dissolve and new forms of conflict emerge, these historical insights become more valuable than ever. By studying how history's greatest commanders overcame seemingly insurmountable odds, future military leaders can develop the tactical acumen, strategic vision, and adaptability needed to secure victory while minimizing the human cost of war.

19
Looking Ahead: Speculations on the Future of Warfare

Having completed our systematic analysis of history's most successful military tactics and the signature strategies of fifteen great commanders, we now shift from historical examination to forward-looking speculation. The change in tone that follows is intentional – moving from the academic rigor required for analyzing proven battlefield successes to a more conversational exploration of emerging technologies and unconventional warfare concepts. While the previous chapters dealt with established facts and time-tested strategies, this section ventures into theoretical territory where formal military doctrine has yet to fully develop.

This departure from strict academic analysis serves a specific purpose. The future of warfare will likely involve technologies and tactics that have no historical precedent – autonomous drone swarms, weather manipulation, magnetic vortex shields, ultra-fast vehicles and other concepts that sound like science fiction today but may become battlefield realities tomorrow.

Traditional military education often struggles to address these emerging possibilities, yet students of warfare need exposure to these concepts to prepare for conflicts that will bear little resemblance to historical battles.

What follows are personal observations and theoretical possibilities based on conversations with researchers and personal experience. While some ideas may seem unconventional or even far-fetched, they represent potential avenues of military development that future commanders should consider.

The informal tone in this section reflects the speculative nature of the content and the difficulty of applying rigorous academic analysis to technologies that are still theoretical or classified. However, these concepts deserve attention because tomorrow's battles may be won or lost based on technologies and tactics that seem impossible today.

Just as Oda Nobunaga's early adoption of firearms revolutionized Japanese warfare, and Grant's use of telegraph and railroads transformed military logistics, future commanders who understand and adapt to emerging technologies will hold decisive advantages. The following discussions, while speculative, aim to prepare military thinkers for possibilities that conventional doctrine has yet to address.

THE DRONE REVOLUTION

If warfare eventually moves to a form in the future dominated by drones of all shapes, sizes and functions, as most everyone believes we will, it will come to the point where drones will eventually be used as autonomous bullets launched en masse. In future warfare, hundreds to thousands of inexpensive deadly drones will be launched simultaneously with the goal of totally destroying targets without leaving anyone alive. If this happens, we will likely see something greater than the WWI level of battlefield carnage. You can hide them and then launch them from anywhere, or even carry them to launch points.

Drones are currently being neutralized with lasers, microwaves and electronic jamming means, but the future seems to be massive simultaneous attacks of thousands of inexpensive, AI-guided autonomous mini- or micro-drones whose sheer numbers might escape these defenses and have the potential to create massive casualties on the scale of WWI frontal assaults despite any counter-measures.

A story comes to mind that might offer some insights into solving this problem. It involves a highly accomplished scientist friend of mine, recently deceased, who told me how to create an electrostatic bubble around yourself. He said you can do this by creating a magnetic vortex that neutralizes any electrical circuits within its range. The magnetic vortex would interrupt the flow of electricity powering electronic circuits within its perimeter, and thus could serve as an electronic off-switch as drones got closer to one's body. Naturally this discussion treads into the territory of speculative science.

He first discovered the method for creating a magnetic vortex when he glued four small bar magnets together in a tight cross shape on a piece of paper. The magnets were glued together with their common North or South poles all touching together at the center of the cross (even though this is highly repulsive). Then he spun the configuration 30,000-50,000 times or more per second in a vacuum. He recounted that the magnetic vortex he thereby created resulted in all of the phenomena one saw in the movie, *Close Encounters of the Third Kind*, when a UFO stopped above Richard Dreyfuss's truck affecting all its electrical systems.

For my friend, first, all electronic circuitry shut down because electricity flow was interrupted in the vicinity of his spinning magnetic vortex. Second, he said there were anti-gravity effects in its immediate area. Third, the rapidly spinning vortex caused a whining sound to originate from some metal objects as the bonds between their atoms were now stretching back and forth thousands of times per second.

Because his job involved inventing the magnetics for satellites (power supply systems), he frequently asked specialized companies to produce magnetic cores for his projects. However, not a single company that he normally contracted would produce a core with the particular composition he specified, which would have enabled him to electronically create this vortex (without need of spinning magnets) so that he could continue to test and refine his discovery. One company told him that they were under government orders to refuse any requests to produce cores whose specifications came within 1% of what he had asked them to create.

This was some forty to fifty years ago – the government was obviously aware of this phenomenon but wanted to keep it secret for various reasons. Since this was decades ago, the related science has no doubt progressed since then. It's now said that a spinning core of mercury vapor, which would produce a magnetic vortex along with the attendant anti-gravity effects, is at the core of the TR-3B black triangular UFO craft that is said to appear in our night skies.

The point is, if his information is true then it would be very easy to use these principles to produce extremely tiny personal devices that could produce a magnetic shield around individuals to render electronic drone attacks useless. You might even be able to produce a ring-sized device that made lifting heavy objects much lighter due to the gravitational effects.

PREDICTING FUTURE CONFLICTS

Financial analyst Martin Armstrong, a well-known advisor to central banks, has developed outstanding computer models that predict events far into the future, and I also created predictive models during my time on Wall Street. As outlined in my book, *The Secret Inner Teachings of Daoism*, one such model predicts a major global conflict – potentially a Third World War – beginning around 2080-81.

America currently has many military technological capabilities that the world has never seen, but by that time, the technology will have advanced considerably beyond even the currently unseen capabilities. Drone warfare, satellite beam weapons, electronic weaponry, superfast air vehicles, and AI-driven autonomous micro-drones will likely be quite common, highly sophisticated and widely deployed.

Ukraine's recent rhetoric labeling certain U.S. citizens "sponsors of war" who should be assassinated because they disagreed with the Ukrainian government, suggests that this type of belligerent behavior will definitely exist in the future among nations, rogue or not, and possibly result in the assassination of influential ordinary citizens who simply disagreed with another government and just off-handedly spoke their mind. One might therefore logically conclude that micro-drones will eventually be used by enemy powers to target generals and politicians responsible for aggression against their countries in some way. Who will vote for war when that happens?

When politicians realize that they would become easy assassination targets just by voting to authorize military action, or for simply speaking out against a foreign nation, and when they realize that they could be easily eliminated by unstoppable micro-drones in response, their enthusiasm for warfare will diminish significantly. With no foolproof protection available, supporting military action (or just opposing a foreign power in some way) could become a death sentence for politicians or free speech, fundamentally changing the calculus behind initiating conflicts. Conventional warfare would cease as all significant leaders behind conflicts (on both sides) would naturally be killed. Drones will not only become more able to reap destruction over time, but can be hidden by enemies anywhere in a country and then launched when desired to destroy almost anything or anyone.

If the international trade (or banking) system returns to gold-backed payments in some form, nations would face challenges financing military operations, too, similar to Britain's difficulties funding WWI under the gold standard, which would also suppress the incidence of wars. The rule is that if people cannot fund wars there won't be wars.

This funding constraint, which would be caused by some form of gold standard for monetary systems, would likely reduce the frequency of wars simply because they could not be easily financed anymore; a nation could not easily inflate its currency or go into debt, and other nations would not readily grant it loans either unless those loans were part of some scheme to gain control of it.

Incidentally, as gold rises in importance going forwards, leading powers will closely examine underdeveloped gold-mining countries, potentially seeking to control their mining output though methods that would essentially disguise a military takeover. This warning especially holds true for many African nations.

Perhaps the forecast of Wernher von Braun will come true that as warfare involving rogue nations declines (because of gold, micro-drones or other reasons) the military industrial complex, in order to sustain itself, will fabricate new "expensive" threats like asteroid impacts or alien invasions to justify on-going military spending.

The ability to assassinate pro-war politicians through untraceable micro-drones, combined with gold-based financial constraints blocking currency printing or the raising of debt, might truly result in the reduction of conventional warfare and put such companies in the red if a solution were not found. In response, defense contractors might pivot toward building government funded planetary defense systems to maintain their operations, though such scenarios are highly speculative. The military-industrial complex needs wars to make money, so if the normal avenues of warfare shut down it must shift to another income stream to survive.

A final forecast arises after seeing the new urban air taxis (i.e., Volcopter, Joby, Archer), the Jetson One personal eVTOL, and the Volonaut flying airbike. Such autonomous vehicles may give birth to "distributed warfare" in the future where powerful but inexpensive weaponized vehicles (including unmanned drones) are independently parked nearly everywhere in a country rather than in a centralized location that everyone knows about. An entire stockpile of weapons can be hidden in this way. Even a nation's enemies can infiltrate that country and pre-position such devices wherever they want, as is the known case for bombs or drones.

Flying drone cars or bikes that carry people (rather than just drones) can be easily weaponized, produced inexpensively, do not need runways or roads or waterways that limit travel routes, and perhaps can eventually become solar-powered, eliminating the need for fuel. Their high mobility would enable a nation to readily protect its territory since they could be located (distributed) nearly everywhere in hidden or overt locations. Every grid on a map could secretly harbor unknown weapon systems.

WEATHER ENGINEERING

Yet another aspect of woo-woo science must be mentioned due to its current usage in warfare, and that's the area of weather modification. Many people remember another scene in *Patton* where the general prayed for the end of rain so that his Third Army could begin to fight in the Battle of the Bulge: "Almighty and most merciful Father, we humbly beseech Thee, of Thy great goodness, to restrain these immoderate rains with which we have had to contend. Grant us fair weather for Battle."

Many in our list of top generals employed nighttime raids and bad weather to launch successful assaults, especially Caesar. To gain control of the weather would be useful in warfare, and also for agriculture. It is already common knowledge that various nations have the ability to affect weather systems, and thus warfare, through microwave and other modalities. There have even been indications that certain countries were threatened with droughts or earthquakes (caused by microwave beaming at critical fault lines) if they didn't do as the great powers sometimes asked. This is blackmail in the threat of weather or earthquake reprisals.

It is very easy to affect the weather on your own. I once saw a Youtube interview where an ex-military intelligence official mentioned that some individual was able to create standing waves that prevented rain in his locale by somehow using a car battery connected to old television antennas, but he didn't explain how the set-up worked. He simply said that the effect was unmistakable on radar weather maps and because it was easily identified, the U.S. government sent some men asking the man to stop.

What I do know something about are old Chinese Daoist methods for creating rain, and modern DIY methods pioneered by Trevor Constable, author of *Loom of the Future: The Weather Engineering Work of Trevor James Constable*, and producer of "Etheric Weathering Engineering on the High Seas" (video), "Aetheric Rainmaking" (video), and his very informative Hawaiian video entitled "Airborne Etheric Rain Engineering Operations."

His two videos are the most enlightening of this series, showing how to create rain using equipment he invented that nearly anyone can duplicate. His books and videos explain how to use very simple-to-construct esoteric devices for weather engineering whereas the U.S. government secretly uses microwave energy and very large-scale atmospheric heating devices such as HAARP. Trevor's equipment costs pennies while the government spends billions.

Trevor taught how to build and use his famous ABS plastic rainmaking tubes that have a double-sided mirror on one end while being wrapped by a thin, highly reflective silver coating. When visiting in Hong Kong, during a trip to China's Hainan island, and during a trip to the Tenggri Desert in Inner Mongolia, my colleagues and I observed that his weather tubes really worked just as demonstrated in his videos.

We witnessed Trevor perform the unprecedented feat of stopping a typhoon in its tracks off of Hong Kong, and holding it there for several days in a stationary position, by using the repulsive power of a simple rain tube pointed in its direction. The mirrored tube also caused an unexpected, strange pink aura in the sky at a 90-degree angle to its pointing direction, which was foretold in his book, *Loom of the Future*.

On Hainan, he created a perfectly circular dark rain cloud by rotating a mirrored rain tube with his hand that eventually caused the cloud that he had formed to burst into showers. He even punched the cloud at its periphery using another tube pointed just inside its circumference that then produced another perfect circle within it. I had never seen anything like this – a perfectly round black cloud directly over-head that was formed by rotating a rain tube, and another smaller but circular empty hole inside it!

He explained in his Hawaiian rainmaking video that his rainmaking tubes create rain by creating a negative Qi-energy pole inside their length. He created a plastic tube that is internally absent of Qi energy because its mirrored surfaces don't allow any Qi to enter inside it through its length. A double-sided mirror fastened on one end of the tube always pushes Qi away so that it does not enter the tube from the outside while the inside side of the mirror pumps out any Qi inside the tube, thus emptying it.

In other words, the design of the tubes reflects away any Qi trying to enter the tube, and the internal mirror inside the tube at one end, reflecting outwards, pushes any internal Qi outwards as per Chinese feng-shui principles. This leaves the tube empty of Qi, and thus it becomes a negative Qi-pole that will automatically grab onto any large bodies of environmental Qi, such as in rain clouds. The mirrors never let any Qi enter or stay inside the tube, because according to feng-shui principles a silver mirror coating reflects away Qi.

By becoming a negative (empty) pole of Qi energy, the tubes can be used to (lock onto and) push against oncoming atmospheric Qi (that has positive energy) until it bunches up into an agglomerated mass and then discharges as rain when the accumulated sky water becomes too large. Trevor demonstrates this in his Hawaiian airborne video.

Other types of tubes can also be used to gradually drag atmospheric phenomena from one place to another, thus changing the long-term weather patterns of smog or rainfall in a region. Trevor has shown in his books how to change long-term regional weather patterns, which can work against any electronically generated weather interference patterns. One need only place the ends of empty pipes (of special dimensions) in a running stream while pointing them correctly in the sky to draw Qi-based rain energy to an area or push it away, and this can also be used to push smog away from a region.

The reason for mentioning these tubes is to offer a counteroffensive to intrusive government weather engineering designed to create droughts in an area to hurt farmers. The tubes, when employed correctly in a rotating circle, can even be used to create fog or lift fog. This simple technology can be used to offset weather warfare, such as the blackmail of drought against a nation's agricultural output in order that it complies with another nation's wishes.

Trevor also directed me to the book, *War and the Weather, or The Artificial Production of Rain,* by Edward Powers (1871). This work documents numerous instances throughout American history where rainfall usually followed heavy artillery fire or gun battles, prompting the U.S. government to commission a study to investigate and explain this phenomenon. The Battle of Fort Donelson, attack on Vicksburg, and many other Civil War engagements cited in the book demonstrated this curious occurrence.

Research revealed that the potassium nitrate (saltpeter) used in American gunpowder production originally came from Chile's Atacama mines. When production declined and manufacturers sourced the chemical from other regions, the rain-inducing effect from burning gunpowder (composed of this particular saltpeter) disappeared. The famous Hatfield brothers, known for their rainmaking abilities, reportedly burned this Chilean potassium nitrate while cooking a brew of other chemicals as a distraction (misdirection), keeping their true rainmaking method secret. Their rainmaking success also diminished as Chilean saltpeter became less available so the brothers were eventually discredited.

Trevor theorized that the Chilean potassium nitrate contained unique trace contaminants that, when burned, created an atmospheric effect triggering rainfall. If researchers could identify these lost contaminants then this could potentially restore the ability to produce rain on demand when someone burns the same concoction. The book *War and the Weather* provides clear documentation of rainfall occurring immediately after American battles involving heavy gunpowder use, and the list of battles and subsequent rainfall is extensive.

ABOUT THE AUTHOR

An ex-Wall Street analyst who has studied world supremacy cycles and prosperity geopolitics, Bill Bodri is the author of a number of non-fiction books in various fields:

The I Ching Revealed: Tap Into the Five Secret Patterns Underlying the 64 Hexagrams – This groundbreaking translation of the I Ching, based on the oldest version found in the Mawangdui manuscript, reveals it as a collection of sixty-four essays on topics of interest rather than a divination manual. Many hexagrams contain commentary on military topics because it focuses on the Zhou dynasty's conquest of the corrupt Shang dynasty of China. Hexagrams 5, 7, 8, 10, 11, 13, 15, 24, 27, 34, 35, 41, 42, 43, 58, 63 and 64 all concern military affairs.

The Art of Political Power: A Translation and Commentary of The Book of Master Guan (The Guan Tzu) – A translation of an important Chinese strategic classic, often considered superior to the *Sun Tzu* because of its higher-level perspective on geo-political moves and political strategies for conquest or leadership rather than just military tactics. Everyone reads *Sun Tzu* but they should read *Guan Tzu (Guanzi)*, too, for elite geopolitical insight into national economic, political and military strategies. It's about structural power theory. The *Guan Tzu* reveals *economic warfare principles* for conquering countries without fighting, and lessons for how ordinary leaders can make themselves into dominant political leaders, as well as how to transform a country into a superpower. An essential read for anyone interested in geopolitics, economics, military strategy, political power, national maneuverings and the hidden forces that shape world events. These lessons are a companion to Sun Tzu's writing on military tactics, but view matters from the higher ruler's perspective.

Culture, Country, City, Company, Person, Purpose, Passion World: The Grand Strategies and Unifying Principles Behind the Groups Which Rise and Thrive – Why do civilizations and countries rise and fall? What principles determine which companies thrive while others collapse? This historical analysis reveals the underlying principles of growth and prosperity ruling various organizational structures (civilizations, empires, countries, cities, companies), and how specific economic cycles which affect these structures can be harnessed by individuals, companies or countries as forces for generational wealth. It contains much of the information people want regarding civilizational and national collapse patterns. Military campaigns can win an empire, but what do you do afterwards? This book tells you, and reveals why we are seeing all the chaos across nations abandoning common sense and suffering monetary, immigration and defense problems at a time when several cyclical change patterns coincide. Its contents will help those who want to deal with Themistocles's objective: "I never learned how to tune a harp, or play upon a lute, but I know how to raise a small and inconsiderable city to glory and greatness."

Neijia Yoga: Nei Gong for Yoga and the Martial Arts – The martial arts emphasize ideal body alignments for physical movements designed to express tremendous muscle power and energy for self-defense and offense. This book will help you reach the highest levels of martial arts using internal energy exercises that lead your internal energy in tune with your movements. You will learn how to amplify Qi through bone, muscles and organs for striking or protection. Within are the special Chinese esoteric techniques for cultivating the Qi (Prana) of your energy meridians that are Qi circulatory pathways; muscle force transmission lines; *bindus, marma* or acupuncture points; appendages such as arms and legs; body cavities and simple body parts (such as the ears, eyes, teeth, penis, and so forth).

Super Investing: 5 Proven Methods for Beating the Market and Retiring Rich – An investment guide revealing five time-tested techniques that have consistently outperformed markets over the past century. These aren't fads – they are enduring wealth models. The book provides methods that have achieved consistent 20-25% annual compounding rates for decades, and are easy to use.

High Yield Investments, Hard Assets and Asset Protection Strategies – This book teaches you how to invest in different types of high-yield investments (many you've never heard of but are easy to buy) – along with strategies for protecting these investments from institutional risks and government monetary changes through hard assets. The major focus, besides hard asset safety, is on creating a bulletproof portfolio of high-yield income that is essentially a cash-compounding machine for consistent passive income.

Bankism: How the Government's Bank-First Policies are Destroying the Nation and How to Survive the Aftermath of a Coming Dollar Collapse – A reset is coming to the world's financial system. Part of the solution will involve the self-funding state bank, such as the Bank of North Dakota, and precious metals in some form or another. The safest course of action for the transitional period ahead is to own physical gold and silver, as well as select cryptocurrencies (perhaps XRP), in order to safely pass through the coming collapse of the dollar, inevitable decline in asset prices when interest rates rise, and destruction of the U.S. economy as various non-dollar payment systems go online that will supplant the dollar and reduce its predominance in world trade. Inside is a history of what brought us to this state of affairs and what to do to move ahead.

Sport Visualization for the Elite Athlete – The secret mental edge used by Olympic champions and professional superstars is visualization practice. This performance-enhancing guide reveals how to harness visualization techniques and mental rehearsals to dramatically improve athletic abilities – whether in competitive sports, martial arts, yoga, or other physical disciplines. This is one of the psychological components that separates world-class performers from the mediocre. Elite athletes use it because visualization works in improving performance. In short, it works!

Quick, Fast, Done: Simple Time Management Secrets From Some of History's Greatest Leaders – This short but remarkable guide on productivity hacks and time management methods offers techniques for accomplishing more in less time so that you will improve your productivity. Time is the battlefield, so stop drowning in tasks and reclaim your life! Learn to get more done in less time, be more productive, and put your life back on track while others remain trapped in cycles of endless busyness and burnout.

Move Forward: Powerful Strategies for Creating Better Outcomes in Life – This isn't about goals. It's about transformation. Inside are proven 1-2-3 strategies for overcoming the flaws of your genetic inheritance, breaking ingrained habits, and changing a "fated astrological fortune" to create radical personal improvement in areas such as health, business, habits, talent development and spiritual growth. These techniques will help you rewrite your life's script by overcoming limitations. You don't need permission. you need. You need strategy and tactics to achieve better life outcomes.